CAMBRIDGE LIBRARY COLLECTION

Books of enduring scholarly value

Botany and Horticulture

Until the nineteenth century, the investigation of natural phenomena, plants and animals was considered either the preserve of elite scholars or a pastime for the leisured upper classes. As increasing academic rigour and systematisation was brought to the study of 'natural history', its subdisciplines were adopted into university curricula, and learned societies (such as the Royal Horticultural Society, founded in 1804) were established to support research in these areas. A related development was strong enthusiasm for exotic garden plants, which resulted in plant collecting expeditions to every corner of the globe, sometimes with tragic consequences. This series includes accounts of some of those expeditions, detailed reference works on the flora of different regions, and practical advice for amateur and professional gardeners.

A Handbook of Plant-Form

Written and richly illustrated by the Derby-born artist Ernest Ellis Clark (1869–1932), this guide was originally published in 1904 to demonstrate the decorative possibilities of certain plants, mainly English wild flowers, to art students sitting examinations in plant drawing and design. Clark emphasises the importance of retaining a certain amount of botanical accuracy and provides examples of the ornamental possibilities of selected plants in various stages of their development. The language employed in describing the plants is not rigorously scientific and may be understood by those with little familiarity with botanical terms (a brief glossary is also provided). By focusing primarily on accurate renderings of the plants, rather than decorative applications, Clark allows the student's originality to remain unaffected by his personal preferences, and in so doing he allows students to adapt his teachings to their particular tastes and styles.

Cambridge University Press has long been a pioneer in the reissuing of out-of-print titles from its own backlist, producing digital reprints of books that are still sought after by scholars and students but could not be reprinted economically using traditional technology. The Cambridge Library Collection extends this activity to a wider range of books which are still of importance to researchers and professionals, either for the source material they contain, or as landmarks in the history of their academic discipline.

Drawing from the world-renowned collections in the Cambridge University Library and other partner libraries, and guided by the advice of experts in each subject area, Cambridge University Press is using state-of-the-art scanning machines in its own Printing House to capture the content of each book selected for inclusion. The files are processed to give a consistently clear, crisp image, and the books finished to the high quality standard for which the Press is recognised around the world. The latest print-on-demand technology ensures that the books will remain available indefinitely, and that orders for single or multiple copies can quickly be supplied.

The Cambridge Library Collection brings back to life books of enduring scholarly value (including out-of-copyright works originally issued by other publishers) across a wide range of disciplines in the humanities and social sciences and in science and technology.

A Handbook
of Plant-Form

*For Students of Design, Art Schools,
Teachers and Amateurs*

ERNEST E. CLARK

CAMBRIDGE
UNIVERSITY PRESS

CAMBRIDGE
UNIVERSITY PRESS

University Printing House, Cambridge, CB2 8BS, United Kingdom

Published in the United States of America by Cambridge University Press, New York

Cambridge University Press is part of the University of Cambridge.
It furthers the University's mission by disseminating knowledge in the pursuit of
education, learning and research at the highest international levels of excellence.

www.cambridge.org
Information on this title: www.cambridge.org/9781108070065

© in this compilation Cambridge University Press 2014

This edition first published 1904
This digitally printed version 2014

ISBN 978-1-108-07006-5 Paperback

A HANDBOOK

OF

PLANT-FORM

A HANDBOOK OF
PLANT-FORM

FOR STUDENTS OF DESIGN, ART SCHOOLS, TEACHERS & AMATEURS

ONE HUNDRED PLATES,
COMPRISING NEARLY 800 ILLUSTRATIONS

Drawn and Described, and with

AN INTRODUCTORY CHAPTER ON DESIGN
AND A GLOSSARY OF BOTANICAL TERMS

BY

ERNEST E. CLARK

Art Master Derby Technical College
National Silver Medallist in Ornament and Design.

"In every object there is inexhaustible meaning; the
eye sees in it what the eye brings means of seeing."

LONDON :

B. T. BATSFORD, 94 HIGH HOLBORN
MCMIV.

· PREFACE ·

N placing this little work before Art Students, I am keenly aware of the difficulties with which the undertaking is beset, but while it cannot rival the more expensive works of a similar nature, I hope it will hold its own in the sphere of more humble productions.

Objection may be taken to it on the debatable ground of the wisdom of placing in the hands of Art Students ready-made diagrams for reference and use in decorative studies. But this objection may be met by a consideration of the difficulties experienced by many young students in obtaining, at any given moment, the right plant, or the information concerning such plant, which is essential in order to make an original drawing. At the same time, it cannot be too frequently urged upon students that the only right way for them is to make their own studies direct from nature. Indeed, one object of this book would be defeated if it were made to take the place of a student's own personal studies, as its aim is rather to direct such studies, so that they may be of practical use from a decorative point of view.

I have refrained from supplementing the plant drawings with examples of their decorative application to given spaces, believing that, had I done so, a check might possibly have been put upon the student's originality, and that any such suggestions best come from the teacher with whom the student is immediately concerned. Not more botanical terms have been made use of than the circumstances warranted, and these will be found explained in the Glossary when the text does not supply the requisite information.

PREFACE.

I have to thank those personal friends who in one way or another have given me assistance in getting together the material contained in the book, and although nearly the whole of the illustrations are from my own drawings from nature, there are one or two which have been made from original drawings kindly lent me for the purpose.

To Mr. Paulson Townsend I am indebted for generous assistance in various ways.

ERNEST E. CLARK.

Derby, *June*, 1904.

LIST OF PLATES.

INTRODUCTION.

IN treating of plant form as a source of inspiration for the designer, one is struck by the abundance and variety of the material; the endless forms and colours of the vegetable world that face him at the outset.

It is not a case of the want of material, but the excess of it, that forms one of the difficulties in dealing with this subject. After having carefully considered the scope and use of the present volume, a careful selection of plant forms, mainly typical of English wild flowers, to be found during Spring, Summer, Autumn, and Winter, has been made with a view to their appropriateness for decorative treatment. One intention of the present book is to form a guide to the Government examinations in memory plant drawing and design.

The requirements of the examiners in the memory plant examination should be thoroughly understood by the student before attempting the paper, as the time allowed does not admit of any superfluous labour.

A careful drawing of a plant, with analytical details, is to be made from memory by the student, before adapting it decoratively to any of the given spaces.

With this object in view, the student should make careful studies from nature of the plants chosen, so that, at any given time, no great difficulty would be experienced in making a drawing from memory. The points to be closely studied in the selected plant are: its decorative possibilities from every point of view; the characteristic growth from the root upwards and outwards; leaf and stalk junctions with the main stem; front, back, and profile of flower and leaf; the growth of buds and their particular forms; all such growths as the calyx of a flower and bracts, should be seized upon and carefully drawn, as they offer great ornamental possibilities for decorative work. The fruit and seed pods of various plants should also be carefully noted, as here again most beautiful forms are abundant; such, for instance, as those to be found in the Poppy, the Nasturtium, the Dandelion, etc. Tendrils of climbing plants and shrubs also offer great scope; and really no more delightful forms are to be found in nature than these tendril forms. They are of great service to the designer in tying and binding together his design, just as they tie, bind and cling to any obstacle in their way in their natural state; but they require very careful drawing and observation in order to give them that nervous, energetic look with which they are so full. A limp-looking tendril should be avoided.

The particular veining of the leaf must also receive attention, as this is a great characteristic of many plants; young leaves and shoots and leaf-buds must, in their turn, be put down in the category of useful forms to be noted and sketched. Some few sections are also necessary as showing the construction of certain parts, such as the stems, showing whether they be triangular or polygonal or square; but anything like a series of botanical sections is not necessary. The section of certain seed-pods and fruits is sometimes useful, such as the Pomegranate or big Poppy-head, and those of the Pea and Bean tribe.

Having made the memory sketch, then, of the chosen plant, the problem will be, to select one of the given spaces, and suitably fill it with a design based upon the particular plant you have selected. There are several ways of commencing a design to fill a given space. First of all, the student should carefully consider the plant in relation to the space itself; some plant forms lend themselves to certain shapes much better than others. One will compose within a rectangle much better than in a square; another in a square better than in a circle, and so on; but much of this depends upon the ingenuity of the student, and no recipes can be given.

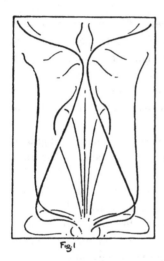 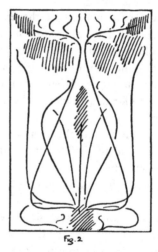 

Fig. 1   Fig. 2   Fig. 3

Having finally settled upon the plant and the given space, a good method is to commence by lightly sketching-in the construction or main lines of the design, getting these to balance each other and to harmonise with the given lines of the space, which are, of course, the controlling lines of the composition. Having decided upon these, suggest with charcoal or soft pencil where the masses or interesting spots will occur, taking care that these also balance; they should generally fall at the most important divisions of the space. Again, the student may begin by placing the masses of the design first, and then connecting them by suitable lines. A third way is, mentally to work out the design before putting pencil to paper; but as this is not usual with students, as it requires long experience to accomplish it, one of the two former ways will be safer and quicker.

The student must fully understand that there is no final word to be spoken regarding the making of decorative compositions. The designer takes that method which comes easiest and most natural to himself; but in most cases it is better and more methodical first of all to decide upon the main lines and masses, satisfying one-self that they balance and harmonise with each other and the lines of the given space.

Having done this, one can then elaborate as much as time and circumstances permit, and good taste and judgment suggest. Figs. 1, 2, and 3, given opposite, will indicate the method of procedure, from the first lines to the finished design.

One or two words might be said regarding the conventionalising of plants. Here, again, at the outset, we are met by an ever-present difficulty in the use of the word *conventionalise*. To begin with, a young student does not understand it in relation to design, and to the older students it presents various meanings. A young student, for instance, does not always grasp the reason why he is not to copy exactly the flower and leaf forms before him, feeling instinctively, and, as a rule, rightly, that they are much more beautiful and ornamental than they will be when he has tortured them into something else, or " conventionalised " them according to instructions. This, of course, is partly true. Therein lies the difficulty, as you cannot conscientiously tell a beginner that his forms *are* an improvement upon nature, even for that particular corner for which he designed them. But what can and must be impressed upon him is, that first of all an ornamental or conventional arrangement of line and mass, based upon the plant, must be decided upon to fill decoratively and harmonise with the given space, before any conventionalising of detail is thought of; in fact, in the process of filling the space the detail is more likely to become satisfactorily dealt with than by what is often no more than a torturing process. To this extent the student will understand that the flower, fruit, or leaf forms of his plant play an important part in the general decorative or *conventional* arrangement of the design, and to that extent are conventionalised or ornamentalised, inasmuch as they take a secondary part in a conventional arrangement. Later he can be initiated, by the assistance of the teacher and examples of conventional forms, into the greater subtleties of and reasons for conventionalising particular flower or fruit and leaf forms. At the commencement it will take all the teacher's energy in preventing his most backward students from making a flower-stem grow in two ways at once, or from placing Wild Roses on a sprig of Holly, and thorns on the Oak. But there are many young students who with very little tuition make excellent designs.

To the student who is more advanced it will not be necessary to quibble over the word at all. He will understand that to take a group of flowers or leaf spray, direct from nature, and place it in a given space without any modification, is neither design nor nature, and a violation of both, because he has not considered his plant in relation to his space, it bears no relation to it whatever, no invention, no artistic feeling was in the least necessary in order to do this. The result is therefore disastrous. But, again, the extent to which an artist may conventionalise natural forms, be they animal or floral, depends *almost* entirely upon the taste of the artist after certain conditions have been fulfilled. I say *almost* advisedly, because he may select a very naturalistic treatment for the decoration of a building, which, in certain parts, would most likely call for a more severe or more conventional treatment. He will understand that after having determined his arrangement with regard to the space at his disposal, the degree to which he carries his conventionalising is left entirely with him—students should not at all be hampered here ; but as a rule they err on the side of naturalness—he may conventionalise his forms out of all semblance to particular natural forms like some of the early Gothic foliage (fig. 4), or he may depart very slightly from the particular

ways and forms of nature, like the decorated Gothic forms (fig. 5).   In deciding this he
is entirely within his own rights, and the resulting general effect either justifies or

Fig.5 Decorated Gothic
14th Cent.

Fig.4 Early English Gothic.
13th Cent.

condemns him. We may say, then, that to conventionalise, is to seize upon the
decorative qualities of growth, flower, fruit, or root forms, and make ornament out of
them, or emphasise them so that they become part of a general harmonious scheme,
ignoring as a rule all natural accidents, such as the number of wrinkles in a Poppy
petal, or the upward or downward turn of a Rose leaf, or the number of revolutions in
a Vegetable Marrow tendril, or in the spiral of a shell.

The process may be further illustrated by the following diagram. Here we have

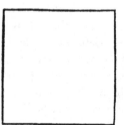

Fig. 6

a given space, say a square, and a
given Rose flower and three Rose
leaves. The problem is to *design*
them into the square so that they
fill it pleasantly, leaving no awkward
spaces. At present it will be seen
that they bear no kind of relation
to each other, the square has no
relation whatever to the floral forms,
or the floral forms to the square ; they are three distinct units from which we are to
make a harmonious composition. To bring them all three into relation—that is, so
that leaves, flower, and square reflect and influence each other—may be done in
somewhat the following manner (*see* fig. 1, pl. 1.).

In this figure we have a simple diagonal arrangement of the three elements
which fall pleasantly in the square form, without violating any principle of growth.
The leaves fit into the corners and arrange themselves along the sides of the square
without any awkward spaces intervening, while the flower is made the centre of
attraction and central *mass* of the design.

Of course, all this is very simple when you are dealing with simple elements,
but exactly the same principles hold when you are dealing with more complex
arrangements. You might, for instance, in this particular case, multiply your
elements by making your space larger, and including the stems, root, bracts, buds, and
fruit of the plant, uniting them all into one arrangement. You have increased your
difficulties by increasing your forms, but not your principles. Similarly, if your design
be for a ceiling, and your elements the human figure and lower forms of animal life,

beyond the designing of the figures and animals to fit into their allotted spaces, the only other great influence on your design would be the materials in which your design was to be executed—whether flat, coloured, or in relief. But this opens up another great question in design which it is not the province of this book to enter into, and which would serve no useful purpose in an elementary treatise.

But to return to our three elements. In fig. 2, pl. 1., we have substituted a circle for our square, thus altering one of our elements. The altered space then brings about an alteration in our design, as the student will see at once that the same design will not do for square and circle alike. In the case of the first the outlines of the square are the controlling lines of the design; in the second case it is the same, but the controlling line this time is the circumference of a circle. The other elements remain the same. The flower remains as before, though not necessarily the central mass, the chief difference being in the arrangement of the leaf forms, which are made to take a somewhat curved line in order to echo the outline of the circle and compose with it. Sufficient has now, I think, been said to show that, no matter what the space is to be filled, the fundamental principles are the same.

There are one or two other principles illustrated in these two little diagrams about which a word or two might be said. Taking the square design again, besides the principle of *subordination* which is illustrated in the fact that the internal arrangement of lines and forms is subordinate to the lines of the space, there is the principle of *radiation* from a centre, illustrated by the leaf-bearing lines running towards the centre of the flower or square. There are several forms of radiation, all of which are to be found throughout nature. An example or two will suffice. If you take an ordinary scallop or cockle shell (fig. 3, pl. 1.), and note the direction of the lines upon it, you will see that they all radiate to one point. This is an instance of radiation from a point or radiation to a point.

The radiation of the main ribs in the leaflets of the Horse Chestnut (fig. 4, pl. 1.), give you the same form of radiation. The radiation of the feathers in a bird's wing is *towards* a given point without actually reaching it (fig. 5, pl. 1.). Then there are radiating tangential lines, either to a straight line or a curved one, sometimes called tangential curvature; this is most generally illustrated by the springing of leaves from a stem or the boughs from a tree trunk—a familiar illustration of it would be the springing of the flower-stalks of the Lily of the Valley (fig. 6, pl. 1.), or the Wild Hyacinth from the main stem—that is, they run down with, and do not cut into, the main stem. This is called radiation to or from a given line. Another principle, which is also illustrated in both the designs, is that of *balance*; but before describing the principle of balance it will be better to give an illustration of the principle of *symmetry*, which is an analogous principle. A symmetrical composition is simply an arrangement whereby the first half of your design is turned over and repeated on the other side of a centre line, real or imaginary, thus making both sides exactly the same, as in the simple example of Greek ornament (fig. 7, pl. 1.), or as you may see in the circle design, where the three leaves on a curved line are repeated exactly or reversed, on the opposite side of an imaginary line. Now if you maintain the balance of line and mass, but slightly alter the detail on one side of your design, you get a balanced arrangement without absolute symmetry, that is you get variety introduced as a new element. The diagram given in fig. 9, pl. 1, of a piece of Renaissance ornament is a very good example of balance. In any design you must have one of these two principles, either absolute symmetry or balance.

The principles of *variety* and *contrast* are also illustrated in figs. 1 and 2, pl. 1., by the pattern occurring as black upon white or *vice versâ*, or by the introduction of a spotted background. Contrast may be obtained in another way—by leaving in your design an empty space upon which the eye may rest, but such space must form a feature of the design and not be merely an accidental vacant space; it must be specially designed like the rest. The little design shown in fig. 10, pl. 1, will explain what is meant, the vacant space in the centre of the panel forming a portion of the design and at the same time a rest for the eye. One other principle might be mentioned, and that is the principle of *proportion*. It is somewhat a matter of personal feeling how far one may go in keeping the proportion between, say the flowers and leaves of one plant in the same design as it exists in nature, or if two plants be used in the same design, how far it is necessary to keep the proportion between them that we find in the plants themselves. If the space to be filled is first divided into separate divisions and two plant forms are to be used, one to be occupied by a design on one flower and the remaining space or spaces by a design on the other, possibly being used as a background to the first, then it is not essential that the strict natural proportion between the two plants should be adhered to; but if the two plants form part of the same design in the same space, then good taste would seem to suggest that their relative natural proportions should be borne in mind. Similarly, with regard to the proportion between the flowers and leaves of the same plant, nature here again will be a safe guide, and a design will generally be of a more refined character where this proportion is observed than where it is ignored. But again, in this case, as in many others, the personal element of the student counts for much, and this is a case where some young students seldom make a mistake, while others require constant watching and warning lest they surround a Hawthorn blossom with leaves of gigantic proportions or a Wild Rose form with microscopic leaves placed on hedgestakes.

There are other principles which it is necessary the student should be familiar with, but which cannot now be entered into here. Sufficient has been said that will, by the aid of the class teacher and by the persistent practice of the student, put them on a fair way to become designers.

First of all, the ability to draw any plant placed before him fairly accurately; then a knowledge of the fundamental principles of design; and lastly, the constant practice of applying the material at hand to spaces of varying shapes, will be all that is necessary for the student at the present stage of design.

The question of the underlying principles of ornament is being further dealt with by the writer in another form, and according to the present syllabus of the Department. Obviously, this is too great a subject to do justice to in the present work, which aims mainly at placing before students material and the manner of collecting material from nature, and certain fundamental rules for its application to decorative purposes.

Plate I.

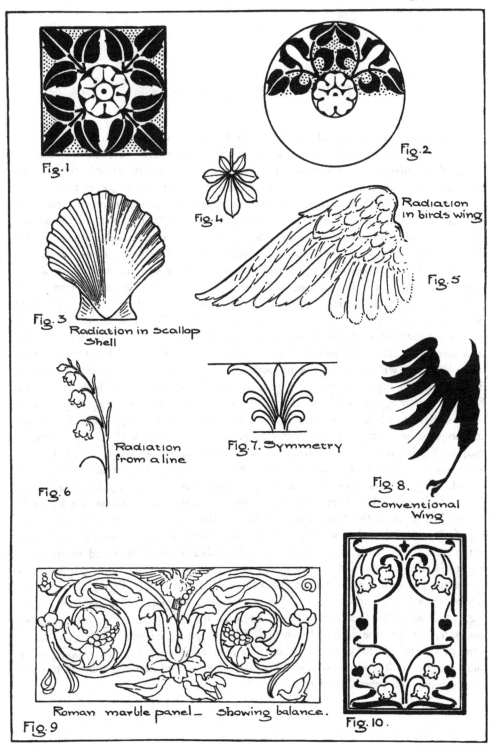

Fig. 1

Fig. 2

Fig. 4

Fig. 3

Radiation in Scallop
Shell

Radiation
in birds wing

Fig. 5

Radiation
from a line

Fig. 7. Symmetry

Fig. 6

Fig. 8.

Conventional
Wing

Roman marble panel — Showing balance.

Fig. 9

Fig. 10.

LEAF FORMS. Plate 2.

On Plate 2 a number of different leaf forms occurring on different plants is given. The variety that exists also in the way they are notched or toothed along their margins is illustrated, and the various ways in which they attach themselves to the main stems. It must be borne in mind that the endless variety of forms to be found in leaves is due to the degree of development to which the leaf has been carried; all leaves, to begin with, are similar in form, but some are arrested in their growth earlier than others, and undergo less modification on that account thus becoming *simple* and *entire* leaves, or other simple forms: degree in development also carries us to *decompound* and *supra-decompound* leaves. The various names attached to the illustrations are explained in the list of botanical terms. A reference to fig. 1, pl. 2, and to the leaf of the Columbine will explain the difference between a *simple* and *entire* leaf and a *decompound* leaf. Fig. 1 is entire because its edge is not notched or divided in any way. Fig. 2 is *orbicular* in form, and its edge is *crenate*. It is also called *peltate*, from the way the stalk is attached to the underside of the leaf, from which point the veins radiate. Fig. 4 is a Wild Rose leaflet, is elliptical in form and its edge is *serrate*. Fig. 4 is *spatulate* in form and likewise serrate. Fig. 5 is *oval* or egg-shaped, and *entire*. Fig. 6 is *ovate* or inversely egg-shaped, with serrated margin. Fig. 7 is another rounded leaf, with serrated edge. Fig. 8 is rather an elongated heart-shape, with a *toothed* edge. Fig. 9 is *oblong*, with a fringed edge or margin. Fig. 10 is the Fir-tree foliage and is called *acicular*, or needle-shaped. Fig. 11 is *cordate* or heart-shaped, and serrated. Fig. 12 is *obcordate*, or inversely heart-shaped. Fig. 13 is *reniform* or kidney-shaped and serrated. Fig. 14 is an example of the *capillary* or hair-like form. Fig. 15 is the *linear* shape, seen also in the leaves of grasses. Fig. 16 is the *linear-lanceolate* leaf of the Willow. Fig. 9a is lanceolate. Fig. 17 is arrow-shaped or saggitate, and is also an example of the sessile attachment to the stem. Fig. 18 is angular in form, and divided or lobed in regard to its edge. Fig. 19 is elliptical also and entire; it also illustrates *parallel* venation as distinct from net-veining, as seen in the leaves of the Ivy or Wild Rose, etc. Fig. 20 is *hastate* or halbert-shaped, and three-lobed. Fig. 21 is strap-shaped or *ligulate*, and entire. Fig. 22 is an *oblique* heart-shaped leaf, such as the Begonia. Fig. 23 is called *runcinate*, owing to the form and direction of its marginal divisions. Fig. 24 shows a form of attachment to the stem which is called *perfoliate*. Fig. 25 *connate*, as in some varieties of the Honeysuckle. Fig. 26 is *amplexicaul*, the base lobes of the leaf passing round and beyond the stem.

Plate 2.

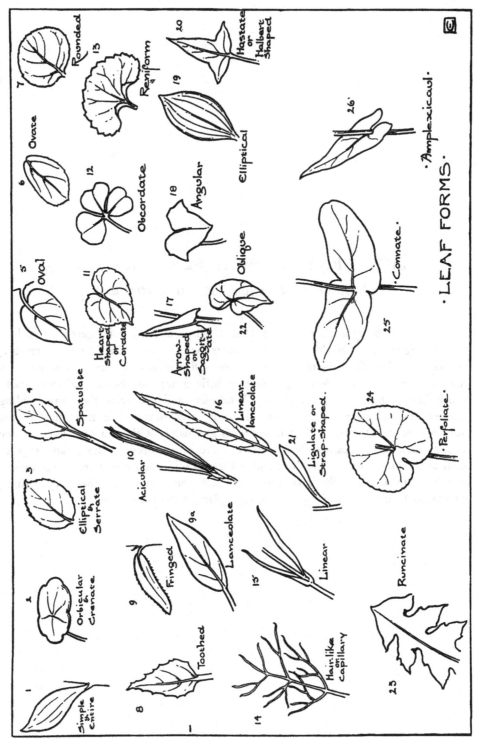

· LEAF FORMS ·

BLACKBERRY, or BRAMBLE. Plates 3, 4, 5.
(*Nat. Ord. Rosaceæ. Rubus fruticosus.*)

THE Blackberry is a conspicuous plant or shrub in English hedgerows, which is rich in suggestions to the designer. Its flowers may be seen from July to September, varying in colour from white to pink. The inflorescence is what is known as a compound panicle; the flowers growing in little groups at the ends of short stalks which branch from the main stem. Petals five and six, with the corresponding number of sepals forming the calyx. Stamens numerous. The fruit may be had from August to October. The leaves grow alternately with, as a rule, five or three leaflets, each on a separate short stalk, and are often lobed. Bracts and stipules grow at the junctions of the flower and leaf stalks with the main stem. One variety of this shrub is called the *Dewberry*, in which the sepals clasp the berry. Buds, flowers, and fruit may be found on the same plant at the same time.

Plate 3

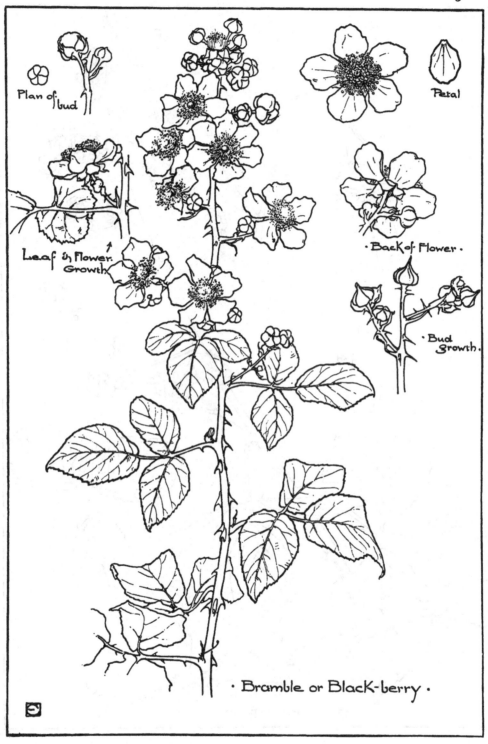

Plan of bud

Petal

Leaf in Flower Growth

Back of Flower

Bud Growth

· Bramble or Black-berry ·

Plate 4.

Fruit Stalks

Leaf Stalk

Back of Berry

Sec.

·BLACKBERRY·

Plate 5.

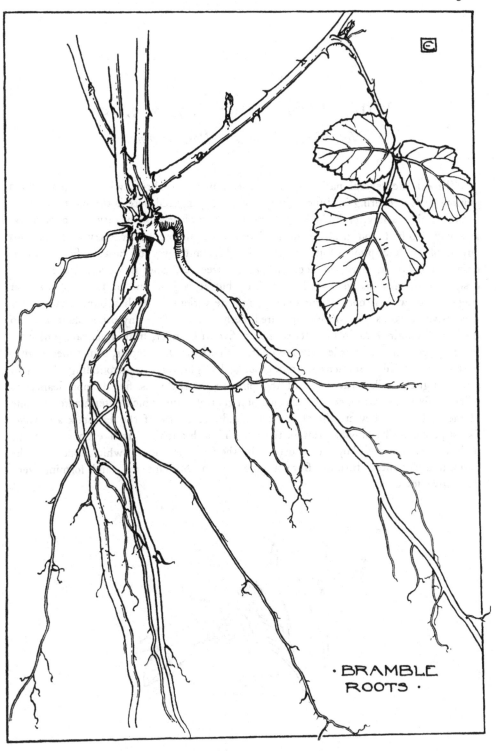

· BRAMBLE
ROOTS ·

WILD ROSE. Plates 6, 7, 8, 9, 10.

(*Nat. Ord. Rosaceæ. Rosa canina.*)

THERE are several native species of this plant or shrub, which forms such a glorious hedge covering in June. Probably the one figured here is the most common, but the white variety, or Field Rose, runs it pretty closely for beauty and popularity, and may also be found in flower much later. It is almost unnecessary to enter into any kind of description of it, as the least observant lover of flowers finds himself drawn towards it, both on account of its sweetness and its lovely form. The designer especially knows it well as a rule; but it is just one of those plants that "age cannot wither, nor custom stale, its infinite variety." It has found material for decorative purposes ever since it appeared, perhaps more than any other plant, and no doubt will continue to do so. Its petals are five in number, from a delicate pink to a much deeper tint; as a rule they are cleft. There are five sepals, sometimes entire, very often divided. Stamens are numerous and bright yellow, surrounding the small sessile stigma. The leaves are of the pinnate order, that is, five or more leaflets of elliptical form, are arranged on each side of a central stem, which is terminated by one of the leaflets. They are sharply serrated, and at the base of the leaf stalk are wing-like stipules. The flowers grow in twos or singly, branching from each side of the main thorny stem. They are succeeded by the fruit or "hips," which appear in the autumn and winter, turning from a green to a brilliant red, and forming very decorative features.

Wild-rose
buds

Plate 6.

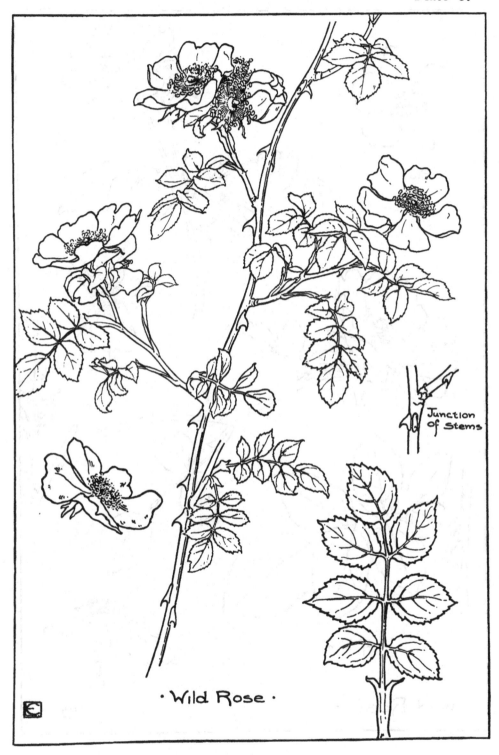

Junction
of Stems

· Wild Rose ·

Plate 7.

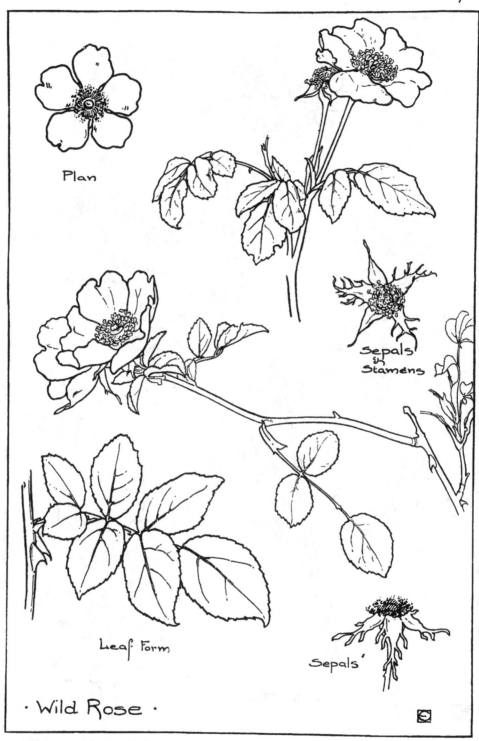

Plan

Sepals & Stamens

Leaf Form

Sepals

· Wild Rose ·

Plate 8.

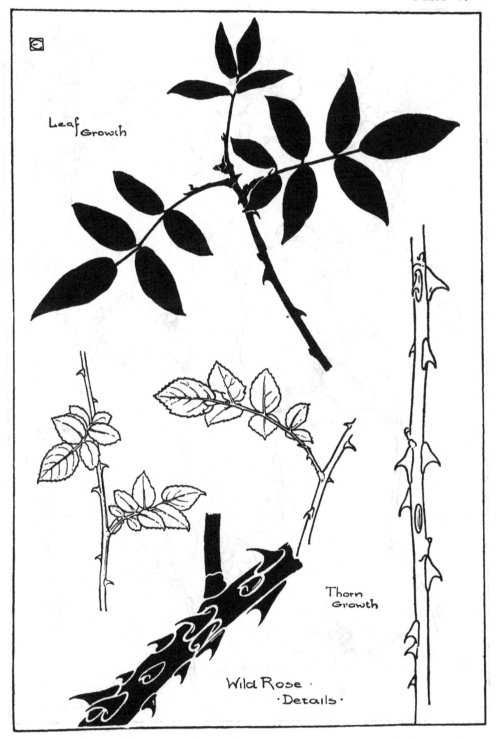

Leaf Growth

Thorn Growth

Wild Rose .
· Details ·

Plate 9.

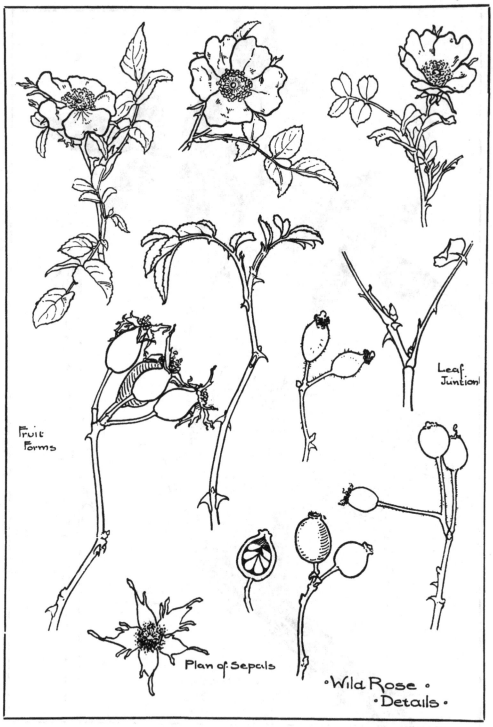

Fruit Forms

Leaf Juntion

Plan of Sepals

Wild Rose Details

Plate 10.

· WILDROSE HIPS ·

THE OX-EYE DAISY. Plate 11.

(*Nat. Ord. Compositæ. Chrysanthemum leucanthemum.*)

THIS is a member of the large *order* of composite flowers ; it assists greatly in turning our mowing grass fields, where it usually grows in groups, into luxuriant flower gardens. It grows from one to two feet in height. The radical leaves or root leaves are stalked and obovate in form, while those from the long flower stems are stalkless or sessile, thinner and coarsely serrated, alternating up the stem. The flower heads are solitary on terminal stems. Petals long and white, notched at the outer extremity, centre yellow. Stems sometimes branched. The involucre consists of a series of green scales overlapping each other and are membranaceous at their edges. This flower is the original of the Chrysanthemum.

Plate 11.

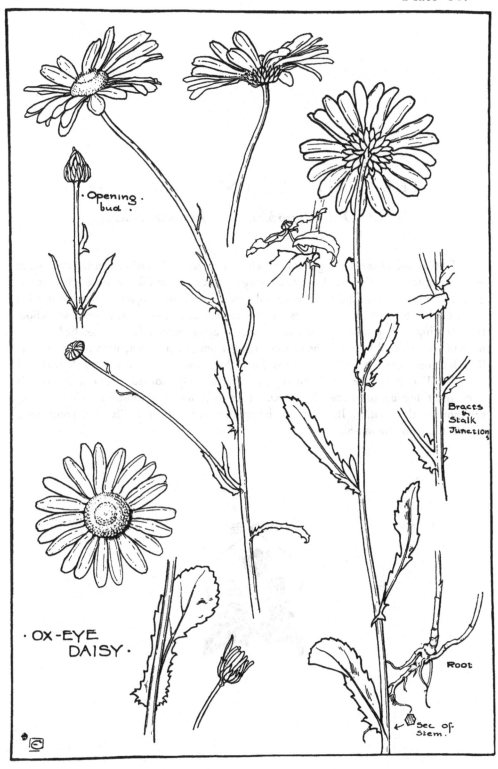

Opening
bud.

OX-EYE
DAISY.

Bracts
&
Stalk
Junction.

Root

Sec of
Stem.

THE WHITE BRYONY. Plates 12, 13.

(*Nat. Ord. Cucurbitaceæ. Bryonia dioica.*)

THIS is one of the trailing plants which makes such a beautiful feature in some of our hedgerows in July and August, forming a thick curtainlike mantle wherever it grows, climbing by the aid of its beautiful tendrils from one object to another until it reaches a height of 6 or 7 feet. Its flowers are small, and of a pale greenish-yellow tint, bearing five petals, five stamens, and a calyx which is five toothed. The flowers are followed by round berries changing from green to red, with a dry surface. The blossoms grow in stalked *racemes* from the same attachment as each leaf and tendril. The leaves are lobed and angular, and coarsely toothed. The root-stock is thick and tuberous, sometimes branched. It is to be found in most English counties, common in the South. It must not be confounded with the Black Bryony, with which it has no relationship.

Plate 12.

·WHITE BRYONY

Plate 13.

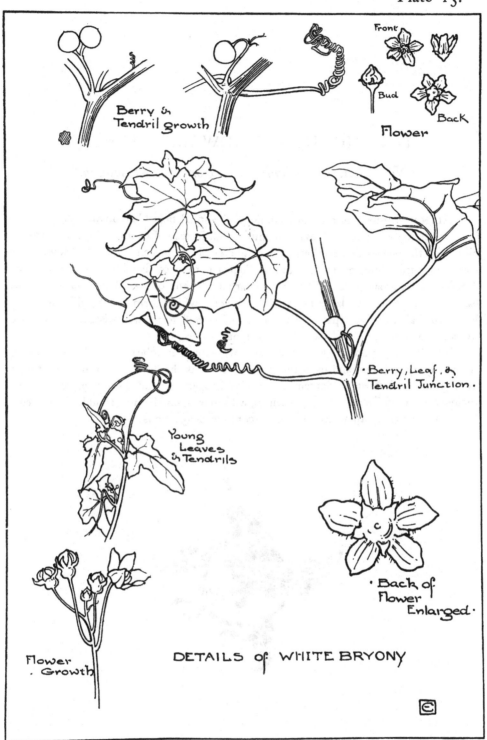

DETAILS of WHITE BRYONY

C 2

THE HOODED BINDWEED.   Plate 14.

(*Nat. Ord. Convolvulaceæ.   Convolvulus sepium.*)

ALTHOUGH this is of the same family as the small field Convolvulus (*C. Arvensis*) it is a much more important-looking member, and, instead of trailing along the ground, climbs up the hedges and thickets to a height of six or seven feet.   Its flowers are large and pure white, having a trumpet-shaped corolla of five lobes or petals, each decorated by a broad plait.   They are solitary at the end of their long stems ; at the base of the tube are two large heart-shaped bracts, which form the hood, completely hiding the calyx of five sepals.   The petals in the bud form are twisted.   There are five stamens, one style, and two oblong stigmas.

The leaves are broad spear-head forms with angular lobes at the base, growing alternately up the stem and from the same junction as the flowers.   It may be found from June to August or September in the hedgerows and bushy places.   The Bindweeds are no friends of the gardeners, who look upon them as troublesome weeds, but to the designer they are rich in decorative suggestions.

Plate 14.

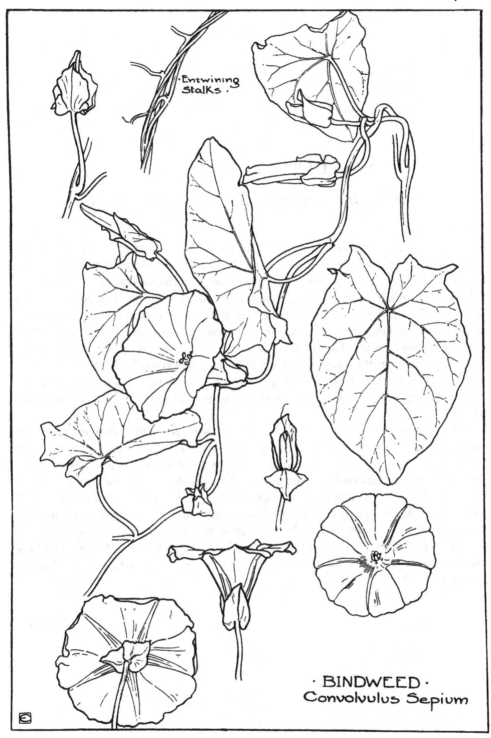

·Entwining
Stalks·

· BINDWEED ·
Convolvulus Sepium

NASTURTIUM. Plates 15, 16.

(*Nat. Ord. Tropæolaceæ. Tropæolum majus.*)

THE common Nasturtium of our gardens, or Indian Cress, as it is sometimes called, is another very useful plant to the designer; its flowers, leaves, and fruit, combined with its trailing growth, offer plenty of material for decorative adaptation. The flowers, which run from a bright yellow to a deep red, have five petals, the two upper ones being slightly larger than the rest, and as a rule are patched and streaked by deep rich markings running from the base upwards; the three lower are fringed near their bases by a hair-like growth. The calyx is one of five divisions terminating in a long spur, from the under-side of which the long flower-stalk grows. The colour of the calyx is a pale yellow generally, sometimes tinged with red; these form the outer covering of the flower-buds, and at this stage take the same pale green colour as the stems.

The leaves are orbicular in form and flat, with a broadly scolloped edge, the stalk is attached near the middle of the leaf on the under side, forming a centre from which the veins radiate to the margin of the leaf. Both leaves and flowers grow from the same junction on both sides of the twisting stalks. The main stem is round and branched. The flowers have eight stamens and a pistil. The fruit which succeeds the flower usually consists of three lobes, which are grooved or channelled. Flowers, buds, and leaves may all be found growing on the same plant together.

The *Tropæolum claratum,* which is figured on Plate 17, is another variety of the same genus. The main difference will be seen in the leaves, which in *T. claratum* have a lobed character.

Plate 15.

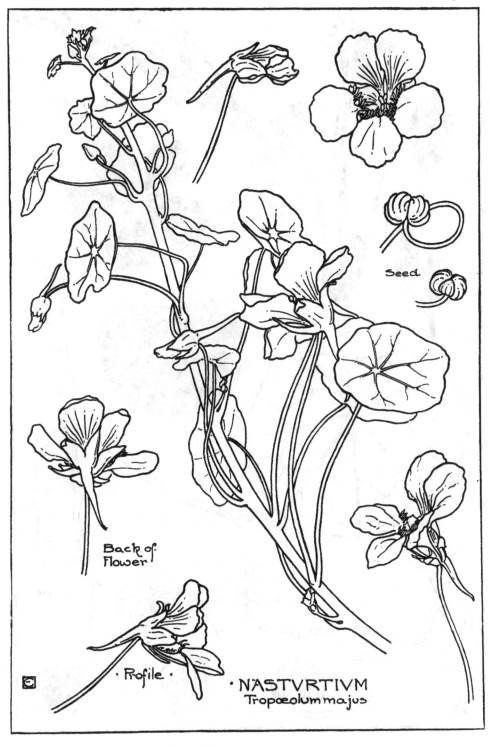

Seed

Back of Flower

· Profile ·

NASTVRTIVM
Tropæolum majus

Plate 16.

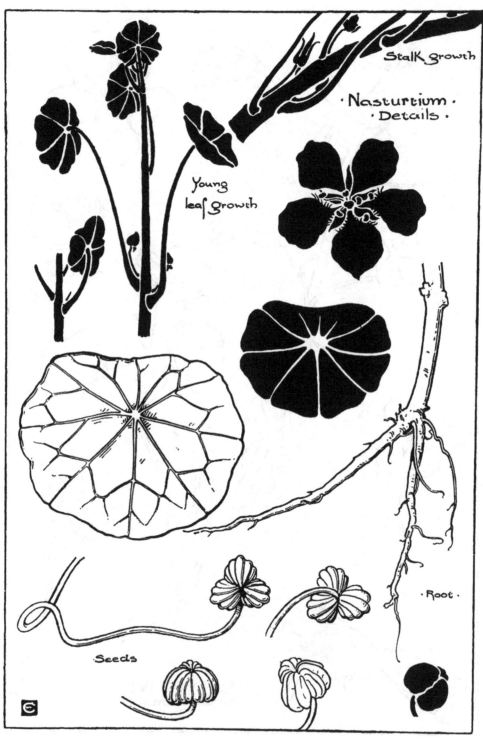

Stalk growth

· Nasturtium ·
· Details ·

Young
leaf growth

· Root ·

Seeds

Plate 17.

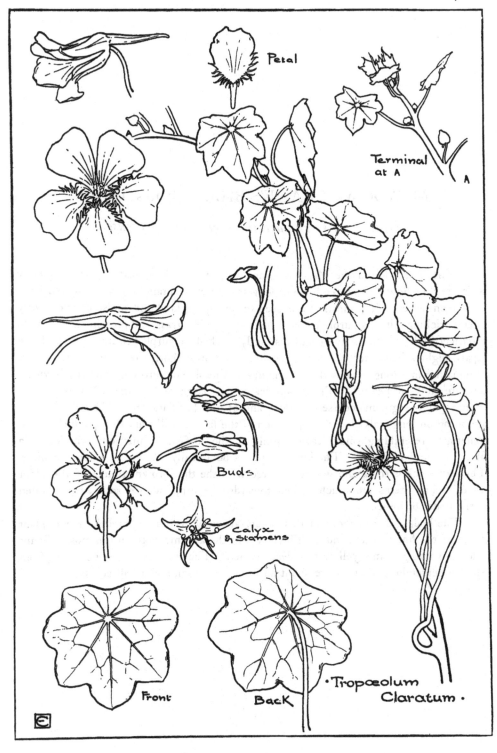

Petal

Terminal at A

A

Buds

Calyx & Stamens

Front

Back

•Tropœolum Claratum•

MEADOW CRANE'S-BILL. Plates 18, 19.

(*Nat. Ord. Geraniaceæ. Geranium pratense.*)

WHERE it is found the Meadow Crane's-Bill is a striking feature, with its large bluish-purple flowers and reddish-purple veins, its erect branching growth, and large dull-green leaves. It is mostly to be seen in low-lying land and moist pastures, very often enriching with its purple blooms the long mowing grass, and growing from one to three feet high. It bears five petals distinctly marked by purple veining, a calyx of five sepals slightly hairy, ten stamens, five long and five short, the ovary is five lobed, terminating in a long beak with five stigmas. The flowers grow in pairs, each with a separate stalk or pedicel. The lower leaves are large and palmate, having five or seven separate segments coarsely serrated and of a rather dark green, the radical leaves are borne on long stalks. At the junction of the flower-stalks are thin pointed bracts, sometimes resembling small leaves, while at the leaf-stalk junctions stipules of a fibrous nature are found. The inflorescence is botanically known as a loose panicle. The root stock is thick and woody, covered with the thin brown stipules of the older leaves. The stems are much forked towards the top, and are somewhat swollen at the nodes or joints.

The buds and fruit forms of this variety are very decorative, and the whole plant is one of the finest of its genus. The leaves in the autumn take on the most brilliant colours, varying from a yellow to a deep scarlet. The flowers are to be found from May to September, and measure about an inch to an inch and a half across.

Plate 18.

· Meadow Crane's-Bill ·

Plate 19.

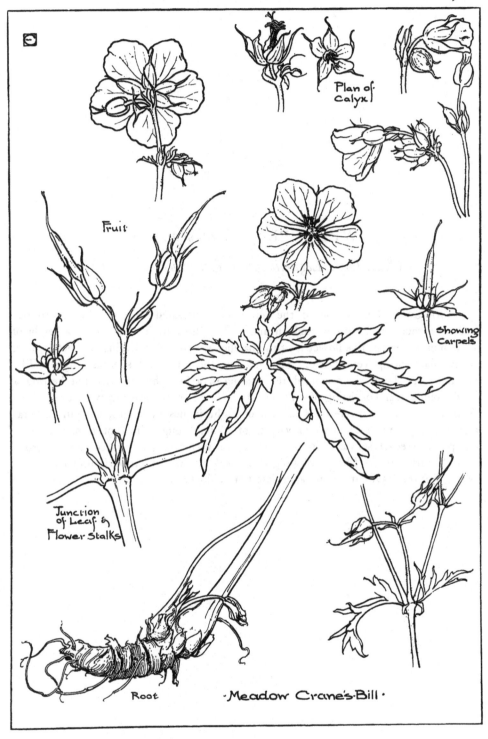

Plan of
Calyx

Fruit

Showing
Carpels

Junction
of Leaf &
Flower stalks

Root

·Meadow Crane's·Bill·

SWEET PEA. Plates 20, 21.

(*Nat. Ord. Leguminosæ. Lathyrus odoratus.*)

THE Sweet Pea is one of the numerous cultivated varieties that adorn our gardens through the summer and autumn. It climbs by its tendrils to a height of seven or eight feet. The flower is of the papilionaceous order, having five petals, one, the standard, being very broad. Stamens ten, the style and stigmas single. The leaf-stalks are flattened or winged, terminating in a branched tendril, and bearing a broad oblong leaf consisting of two leaflets. The stipules are arrow-shaped. The peduncles or main flower-stems are six or eight inches in length, carrying several large flowers, the colours of which vary from white, through pinks, purples, and blues. The pod or seed-vessel varies from one to several inches long ; the seeds are numerous and rather flattened. The Everlasting Pea (*Lathyrus latifolius*) is very similar in every respect, but the leaves are more lanceolate than oblong.

Plate 20.

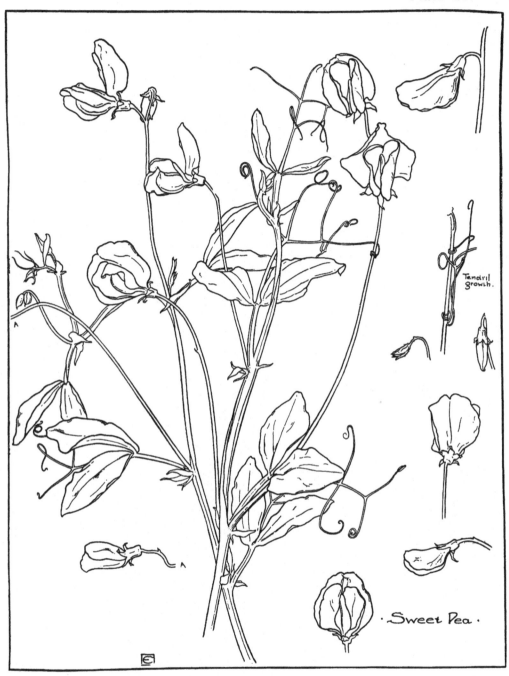

Tendril growth.

· Sweet Pea ·

Plate 21.

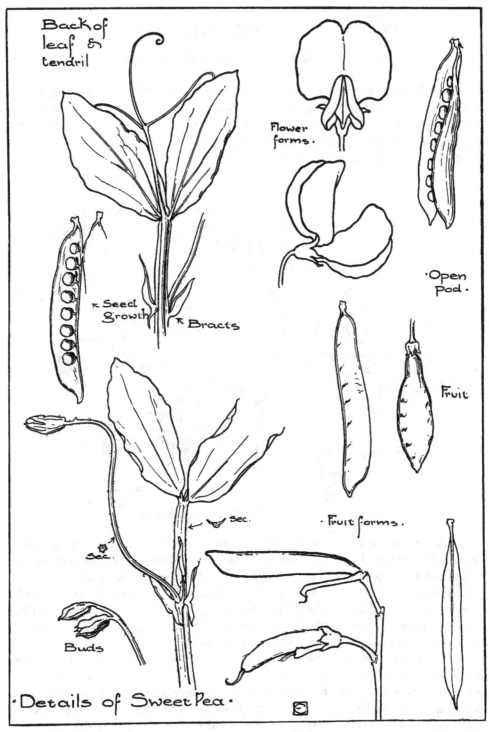

Back of
leaf &
tendril

Flower
forms.

·Open
Pod·

⌐ Seed
Growth   ⌐ Bracts

Fruit

Sec.

· Fruit forms.

Sec.

Buds

·Details of Sweet Pea·

WOODY NIGHTSHADE. Plate 22.

(*Nat. Ord. Solanaceæ. Solanum Dulcamara.*)

THE Nightshade, or Bitter-sweet, is one of those trailing or climbing plants which, not being self-supporting, leans and crawls over the thick vegetation of the hedgerows.   It is most conspicuous when bearing fruit in the autumn, its berries being egg-shaped or oval, turning from a bright green to a bright red, and growing in bunches with a very decorative growth.

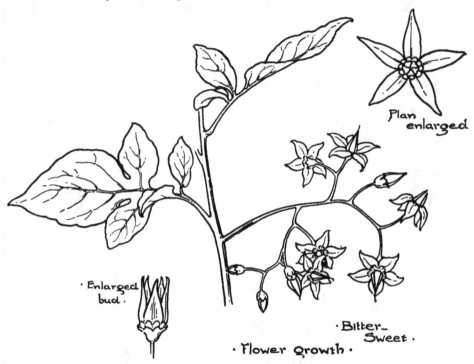

Plan enlarged

· Enlarged bud ·

· Bitter-Sweet ·

· Flower Growth ·

The flowers are small and of a purple colour, having five pointed petals, which are usually reflexed, *i.e.* turned backwards on to the flower stalk, the anthers are united and form a pointed yellow centre, the calyx has five divisions; there is also a small green spot at the base of each petal.   There are five stamens and a pistil. The leaves vary in shape considerably, commencing at the lower part of the stem by being heart-shaped; they gradually get more lanceolate, having often a pair of wing-like lobes at the base, giving the leaf the hastate form.

The groups of flowers grow as a rule from the opposite side of the stem to the leaves, but from the same point on the stem.   This fact makes it rather difficult to get a satisfactory view of both leaves and flowers or berries from the same piece.   Its flowering period is a long one, being from June to September, so that both flowers and berries may be found growing together on the same plant, as is the case with the Blackberry.

Plate 22.

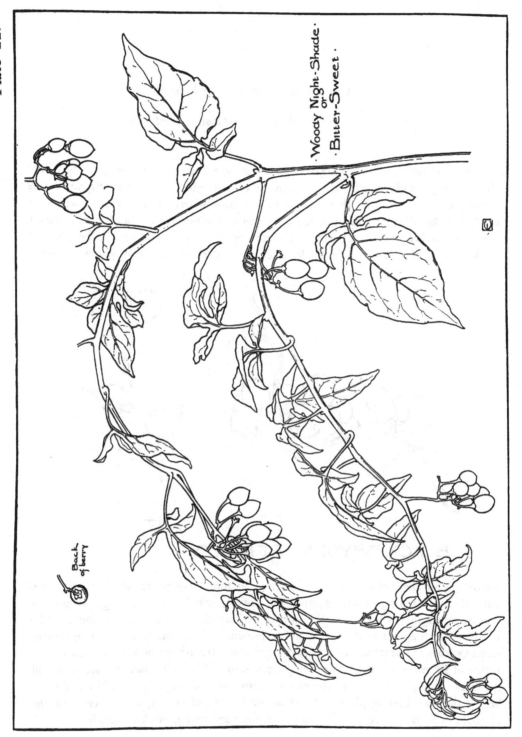

Woody Night-Shade.
or
Bitter-Sweet.

Back of berry

BLACK BRYONY. Plate 23.

(Nat. Ord. Dioscoreaceæ. Tamus communis.)

ALTHOUGH this is a trailing or climbing plant it bears no relationship to the White Bryony (*Bryonia dioica*), which belongs to the Gourd tribe. The Black Bryony forms a conspicuous feature in those southern counties where it is common, when its clusters of large bright green and red berries in the autumn give patches of brilliant colour to the hedgerows. It climbs to some considerable height up the hedges and thickets, draping them with a thick covering of dark-green glossy leaves, growing

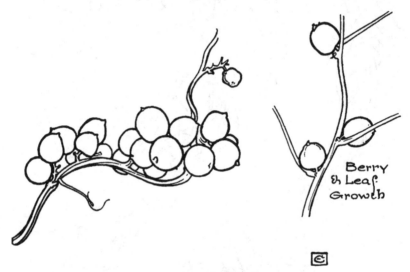

Berry & Leaf Growth

BLACK·BRYONY BERRIES·

alternately up the stem, cordate and pointed in form, sometimes lobed at the base, but otherwise entire, gradating from the lowest, which are very large, sometimes measuring six inches across, to the upper ones, which are less than an inch. The flowers, which appear in the spring and early summer, are small, and grow in slender racemes from the junctions of the leaves, and are yellowish-green in colour, each consisting of six segments, the flower being a perianth. The males have six stamens, and the females the pistil with three stigmas, but grow on separate plants. The one figured here is the male-bearing plant. The fruit or berries, of course, only succeed to the pistillate or female flowers. The roots are large and fleshy, black externally.

Plate 23.

Berry
Growth

Flower

· BLACK BRYONY ·

HAWTHORN (May or Whitethorn). Plates 24, 25.

(*Nat. Ord. Rosaceæ. Cratægus oxyacantha.*)

In May the hedgerows are white with this hedge shrub. This is a plant the designer need feel no superstition about. Its groups of white-and-pink blossoms are not only attractive, by their particular form and sweet odour, to the ordinary flower lover, but by their decorative possibilities to the artist also. The flowers grow in the form of what is botanically known as a *corymb;* they are five-petalled, with calyx of five divisions, stamens many.

The leaves are divided into three and five lobes, deeply serrated as a rule. Stipules large and leafy. The fruit or haws are found through the autumn, forming groups of red ovate berries. Flowers May to middle of June.

Plate 24.

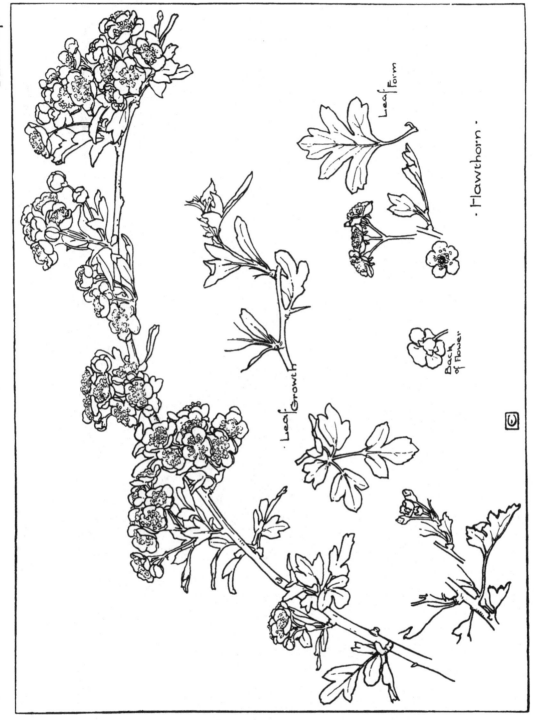

Leaf Form

· Hawthorn ·

Leaf Growth

Back of Flower

Plate 25.

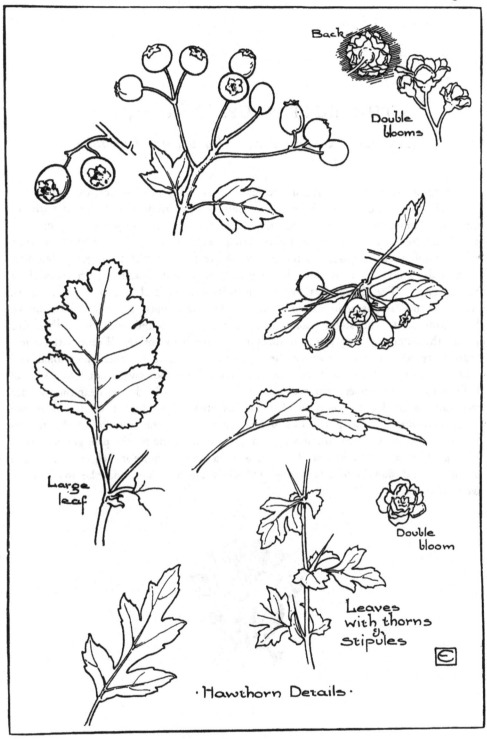

Back

Double blooms

Large leaf

Double bloom

Leaves with thorns & Stipules

·Hawthorn Details·

THE BUTTERCUP.   Plates 26, 27.

(*Nat. Ord. Ranunculaceæ.   Ranunculus bulbosus.*)

THERE are some half-dozen species or this plant, all of which are common in England.   The one figured here is the bulbous Crowfoot, which, along with the meadow Ranunculus, is perhaps the most abundant in our meadows.   Designers as well as landscape artists should feel particularly drawn towards this herald of summer, as it not only turns our pastures and meadows into fields of " cloth of gold," but also, from a decorative point of view, has a fine growth, and contains much material for the designer.   The flower has five bright yellow petals and five pale green sepals, which, in this case, as soon as the flower is fully developed, are reflexed or thrown backwards on to the stem.   Sometimes they may be found with more than five petals, the author recently found a specimen on the banks of the Trent with no less than thirty petals; it is shown in the plate, but of course is quite exceptional, and must therefore not be taken as representative.   The root swells out into a kind of bulb, which is the chief characteristic of this variety.   Its leaves near the root are large and many lobed, deeply divided and serrated; those higher up the stem are longer and thinner.   Long leaf-like bracts are found at the junction of the flower-stalks.   The stamens are numerous and yellow, surrounding the pale green stigma. The plant is erect and branched, growing to a foot or more in height.   It is one of the typical flowers of late spring and early summer, and may be found from May to July.

Plate 26.

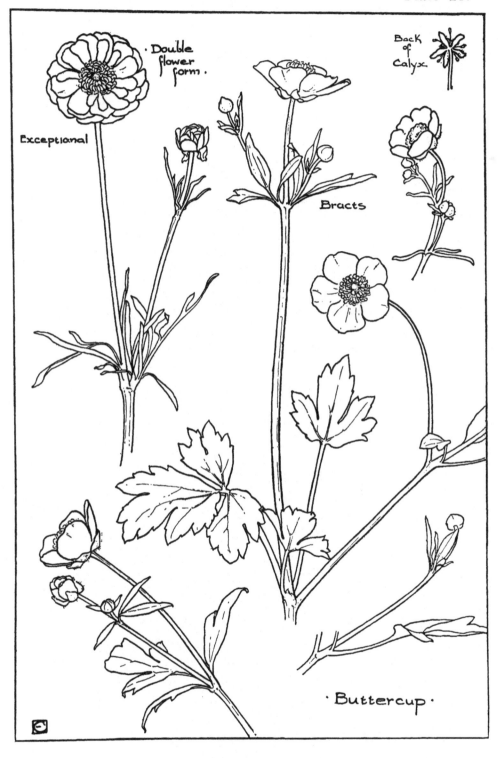

Double
flower
form.

Exceptional

Back
of
Calyx

Bracts

· Buttercup ·

Plate 27.

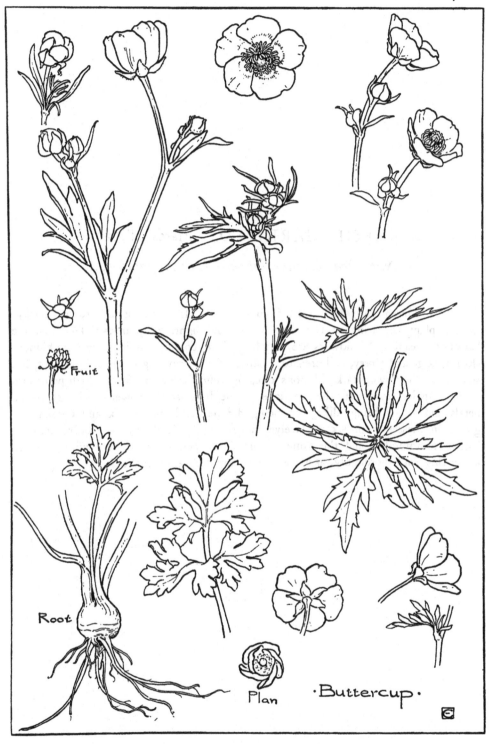

Fruit

Root

Plan

·Buttercup·

MARSH MARIGOLD. Plates 28, 29.

(*Nat. Ord. Ranunculaceæ. Caltha palustris.*)

THE Marsh-marigold, *alias* Mayblob, *alias* Mary-bud, is a strong healthy-looking plant, appearing in the early spring and lighting up marshy land and the banks of streams. Its flowers are a rich gold in colour, and its leaves are kidney-shaped, large and glossy, and crenate. Most of the leaves grow from the root, but some from the flower-stalk junctions clinging round the stem, from which point also fibrous stipules grow. In the language of the botanist the flower is a perianth, the petals and sepals being alike, not being differentiated into calyx and corolla. It grows in masses from six to twelve inches high. Petals five or more, stamens indefinite in number, and the same colour as the petals. Designers will do well to know this plant thoroughly, as it possesses many decorative qualities.

Plate 28.

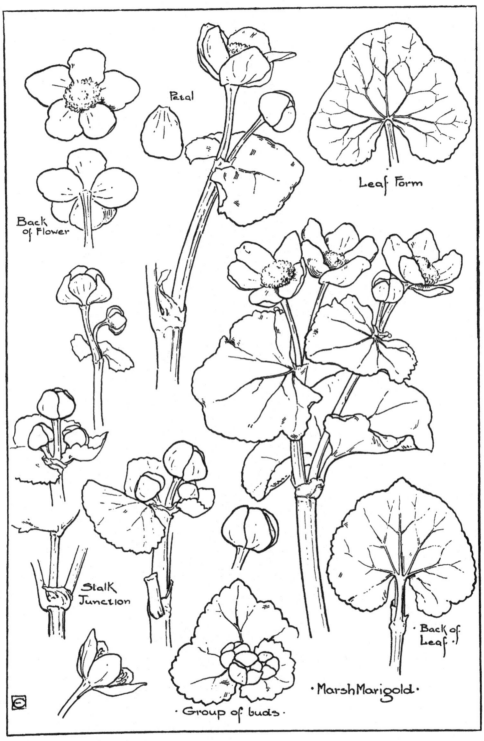

Petal

Leaf Form

Back
of Flower

Stalk
Junction

Back of
Leaf

· Group of buds ·

· MarshMarigold ·

F

Plate 29.

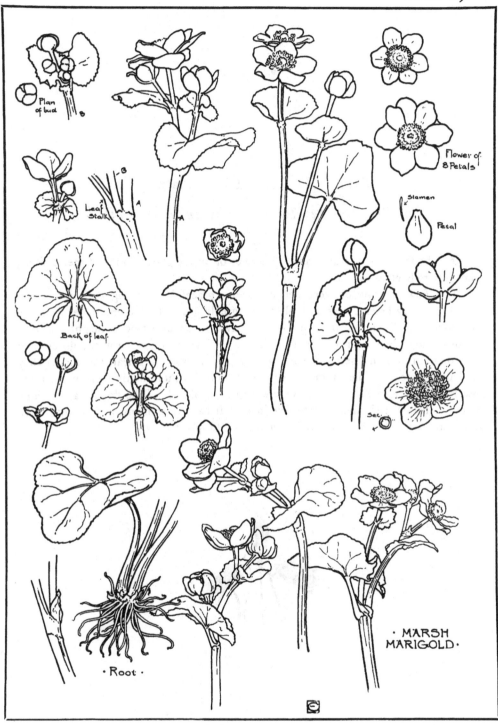

Plan of bud

Leaf Stalk

Back of leaf

Flower of 8 Petals

Stamen

Petal

Sec.

· Root ·

· MARSH MARIGOLD ·

E 2

THE HAREBELL. Plate 30.

(*Nat. Ord. Campanulaceæ. Campanula rotundifolia.*)

Is to be found in pastures and moorland, or in the grassy banks of the roadside, from July to September.

The flowers grow on very delicate stems, gracefully hanging their heads and nodding to every breeze. Several blossoms usually grow at different angles from the main stem, which reaches from an inch or two to about a foot in height. A pale purplish blue is the colour of the flowers. The lowest leaves near the ground are very round in form and serrated, getting long and narrow or linear the higher they reach up the stem. The stem is very often branched, terminating in one or more flowers. The segments of the calyx are long and pointed. The flower, as its name implies, is bell-shaped, divided into five lobes. It has five stamens, a style with three to five stigmas. To the designer it is a very useful plant.

Plate 30.

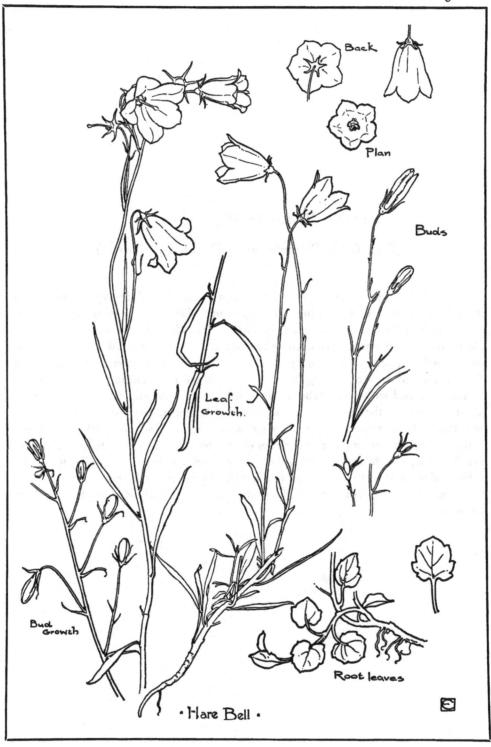

Back

Plan

Buds

Leaf
Growth.

Bud
Growth

Root leaves

· Hare Bell ·

THE PRIMROSE. Plate 31.

(*Nat. Ord. Primulaceæ. Primula vulgaris.*)

THE Primrose is a general favourite among lovers of wild flowers, partly due to its early appearance in the spring. In April and May it can be found in profusion along the hedgebanks, in copses and pastures. Its pale yellow blossoms, measuring about an inch in diameter, grow singly on long radical, leafless stalks. The calyx is slightly inflated, and has five angular divisions of a pale milky green. The petals are five in number, each one being notched or cleft on the margin. There are five stamens enclosed within the tube of the corolla, a style and stigma. The leaves are about as long as the flower-stalks, and of a pale green, obovate in form, with a wrinkled surface, crenate along the margin and radical. The root-stock is thick and fleshy. There are several native varieties of this plant, such as the Oxlip, the Bird's-eye Primrose, and the Scottish Primrose. One allied species, the Cowslip, forms the subject of the next plate. It grows to a height of three to six inches.

The designer, unlike Peter Bell, should find in it something more than a yellow Primrose.

Plate 31.

·PRIMROSE·

THE COWSLIP. Plate 32.

(*Nat. Ord. Primulaceæ. Primula veris.*)

THE Cowslip, or Paigle, is another great favourite wild flower, and bears a close relationship to the Primrose. Although the inflorescence is in both cases an *umbel*, the arrangement of the blooms in the Cowslip is so distinctive as to make it appear a widely different plant. They grow on short stalks from the same point at the extremity of the long radical flower stalk ; from this point also two small bracts grow.

There are five small petals joining together to form a tube, which is mainly hidden by the calyx. The colour of the blooms is from a pale to a deep yellow, with sometimes a bright orange spot at the base of the petals. There are five stamens, and a pistil within the tube. The calyx is five pointed, resembling that of the Primrose.

The leaves are radical, wrinkled and crenate along the margin, and pale green. It grows from three to six inches high in fields and pastures and hedgebanks. Its time of flowering is early spring, as in the case of the previous flower. Very characteristic is the way in which the blossoms of the Cowslip hang their heads.

"Cowslips wan that hang the pensive head."

Plate 32.

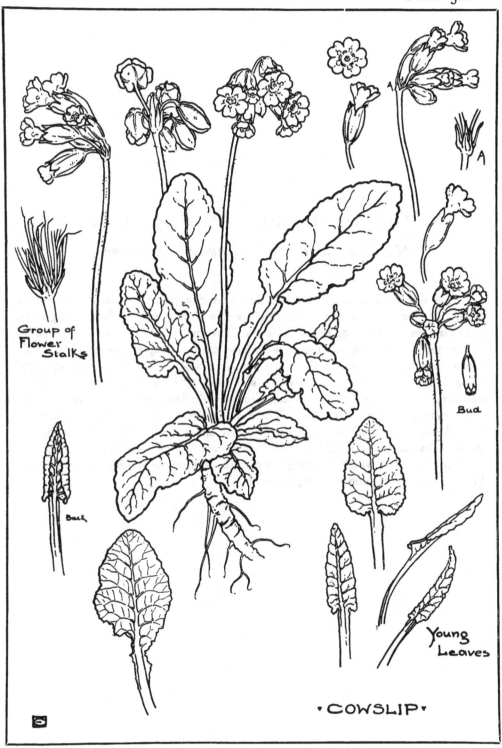

Group of Flower Stalks

Back

A

Bud

Young Leaves

·COWSLIP·

THE DANDELION. Plates 33, 34, 35.

(*Nat. Ord. Compositæ. Leontodon Taraxacum.*)

FEW flowers offer more material for decorative purposes than " this vulgar Herb," the homely Dandelion. The bright yellow flowers may be found the greater part of the year round along the roadside and on waste land, as well as in more culti-vated spots. It is a composite flower, with long, narrow, strap-shaped petals, toothed. When these decay the inner scales of the involucre close and the seed ripens, opening out again into the well-known fairy-like seed-head, which is scattered by the wind broadcast. The leaves and flower-stalks grow straight from the root, the flower-stems being from six to eight inches long. There is considerable difference in the form of the leaf, some being much more deeply lobed and notched than others. Stalks hollow. Stamens five, one style, stigma cleft in two, to each floret. The leaf is, with regard to the way in which the margin is divided, an example of the *runcinate* form. The root-stock terminates in a long tap-root.

Plate 33.

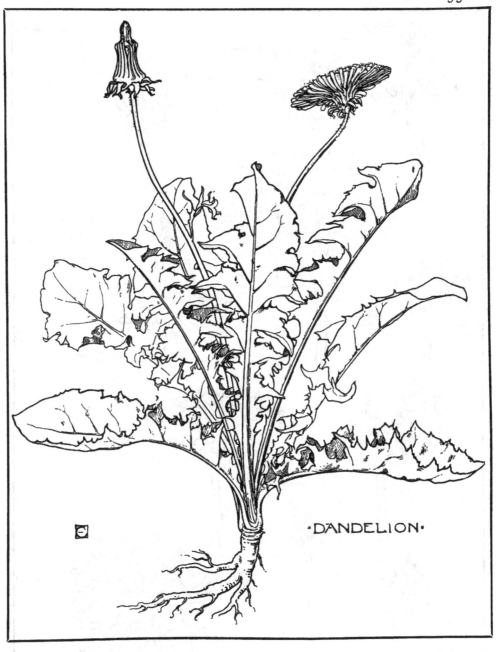

·DANDELION·

Plate 34.

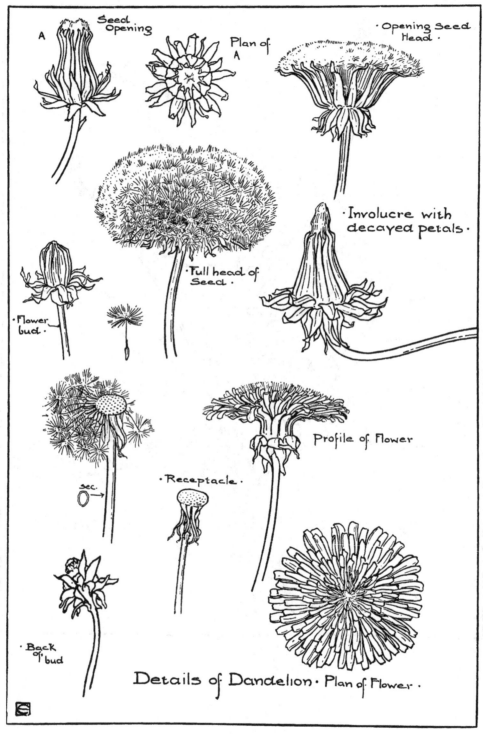

Seed Opening

A

Plan of A

· Opening Seed Head ·

· Involucre with decayed petals ·

· Full head of Seed ·

· Flower bud ·

sec.

· Receptacle ·

Profile of Flower

· Back of bud

Details of Dandelion · Plan of Flower ·

Plate 35.

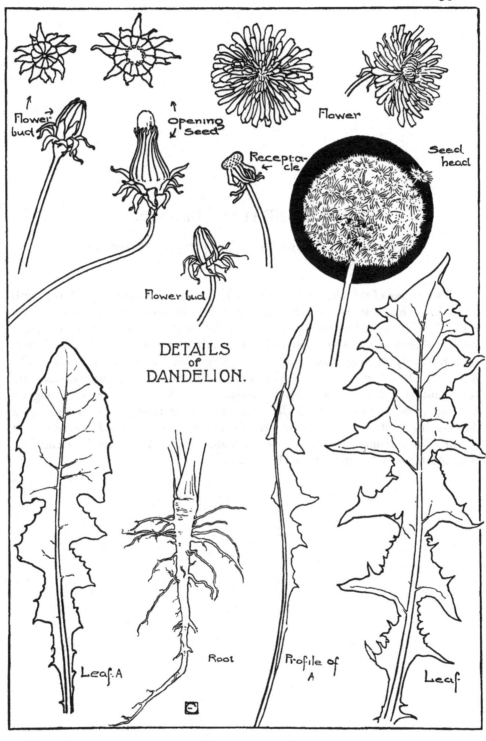

Flower bud

Opening Seed

Flower

Recepta-cle

Seed head

Flower bud

DETAILS
of
DANDELION.

Leaf. A

Root

Profile of A

Leaf

GOAT'S-BEARD.   Plate 36.

(*Nat. Ord. Compositæ.   Tragopogon pratensis.*)

THE flower of this plant is very much like that of the Dandelion, but the sepals or involucral bracts are very much longer, measuring from one inch to an inch and a half.   The leaves are long and grass-like, attached by their bases to the flower-stalk; the radical leaves grow from a few inches to nearly a foot in length.   The flowers are borne singly on long stalks, and are a bright yellow.   The seed-head is also like that of the Dandelion, but larger and coarser.   The buds and opening seed-heads are very beautiful in form, and will repay investigation.   The plant is erect and slightly branching, from one to two feet high, and having a long tap-root.   It is also called the Meadow Salsify, and from its early retiring propensities has earned the somewhat clumsy, if picturesque, name of John-go-to-bed-at-noon.   It flowers June and July in meadows and rich pastures throughout Europe.

Plate 36.

Opening
Seed heads

Root

·Goat's Beard·

THE DAFFODIL.   Plate 37.

(Nat. Ora. Amaryllidaceæ.   Narcissus Pseudo-Narcissus.)

THE Daffodil, long a favourite with the poets, is fast becoming a favourite with designers, and for good reasons.   Its beautiful colour, form, and simple decorative qualities render it specially adapted for simple and effective decorative arrangements.

The flower is a perianth and tubular, having six segments or sepals and petals oblong, and of a more delicate and paler yellow than the crown or bell, which is usually much more golden in colour, the edge of it being irregularly cut or waved. The six stamens spring from the sides of the tube.   The pedicel or flower-stalk is short, and springs from a membranous sheath or spathe.   The flowers are solitary and terminal.   It has a style and stigma which is three-lobed.   The fruit is a capsule. The leaves are long and linear, dark green and radical, but a little shorter than the flower-stalks, which are hollow.   The root is a bulb.   It is to be found in moist woods and copses in March and April.

Plate 37.

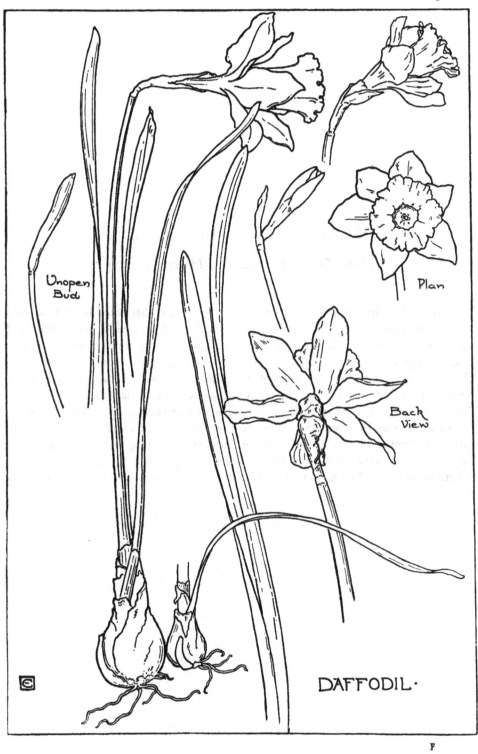

Unopen Bud

Plan

Back View

DAFFODIL.

COMMON IVY.   Plates 38, 39, 40.

(*Nat. Ord. Araliaceæ.   Hedera helix.*)

OF all the climbing plants perhaps the Ivy is the last that could be spared.   We are not only bound to it by many ties and associations of a more or less sentimental nature, but it appeals to us also as a great decorator of our ruined shrines.   It was held in high esteem by the Romans, who weaved it into the poet's crown.   The Greek and Roman artists also selected it as a plant suitable for decorating in relief their marble columns and urns, etc.   It belongs, of course, to the family of Climbers. The flower consists of five petals, and a calyx of five short teeth.   They grow in umbels at the ends of branches; there are five stamens and a short style.   Flowers appear in October, but the fruit is not ripe until the following spring.   The leaves vary in outline very much, sometimes being deeply lobed or five-pointed, and sometimes simple.   The natural order of their growth on the stem is alternate.

Plate 38.

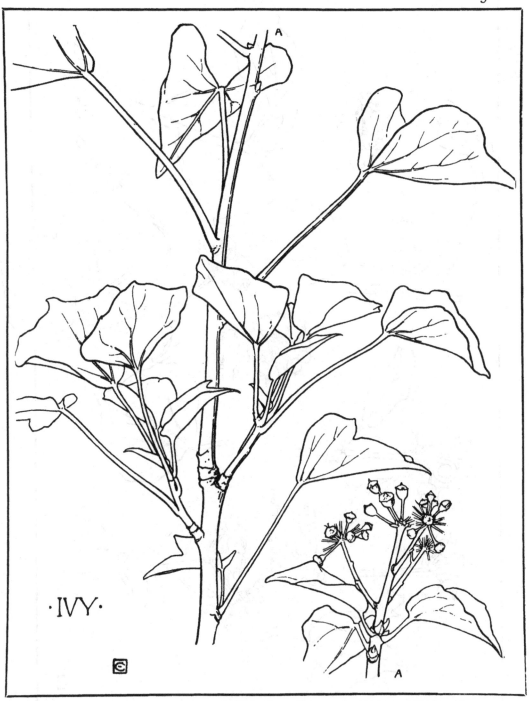

·IVY·

Plate 39.

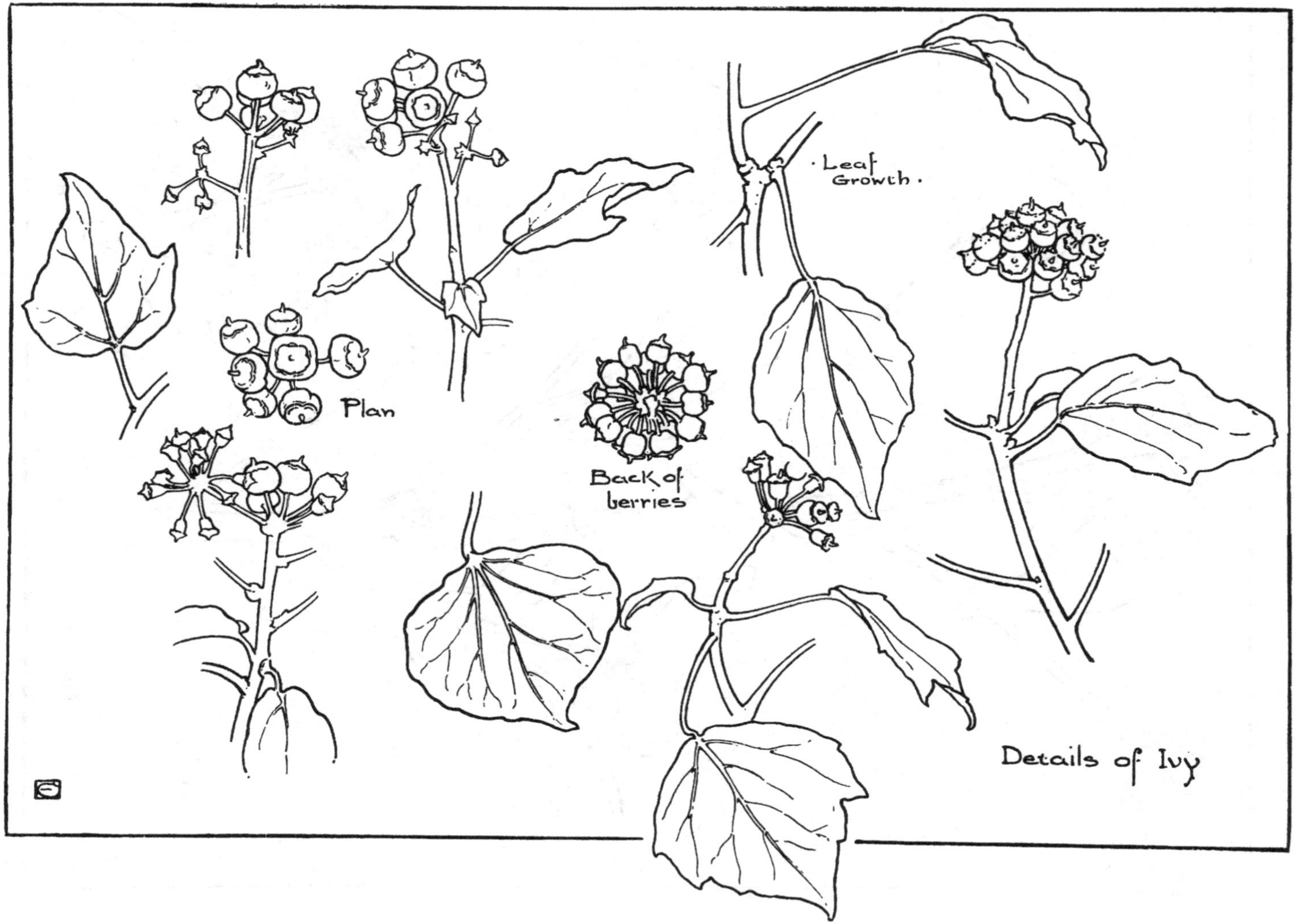

Details of Ivy

. Leaf Growth.

Back of berries

Plan

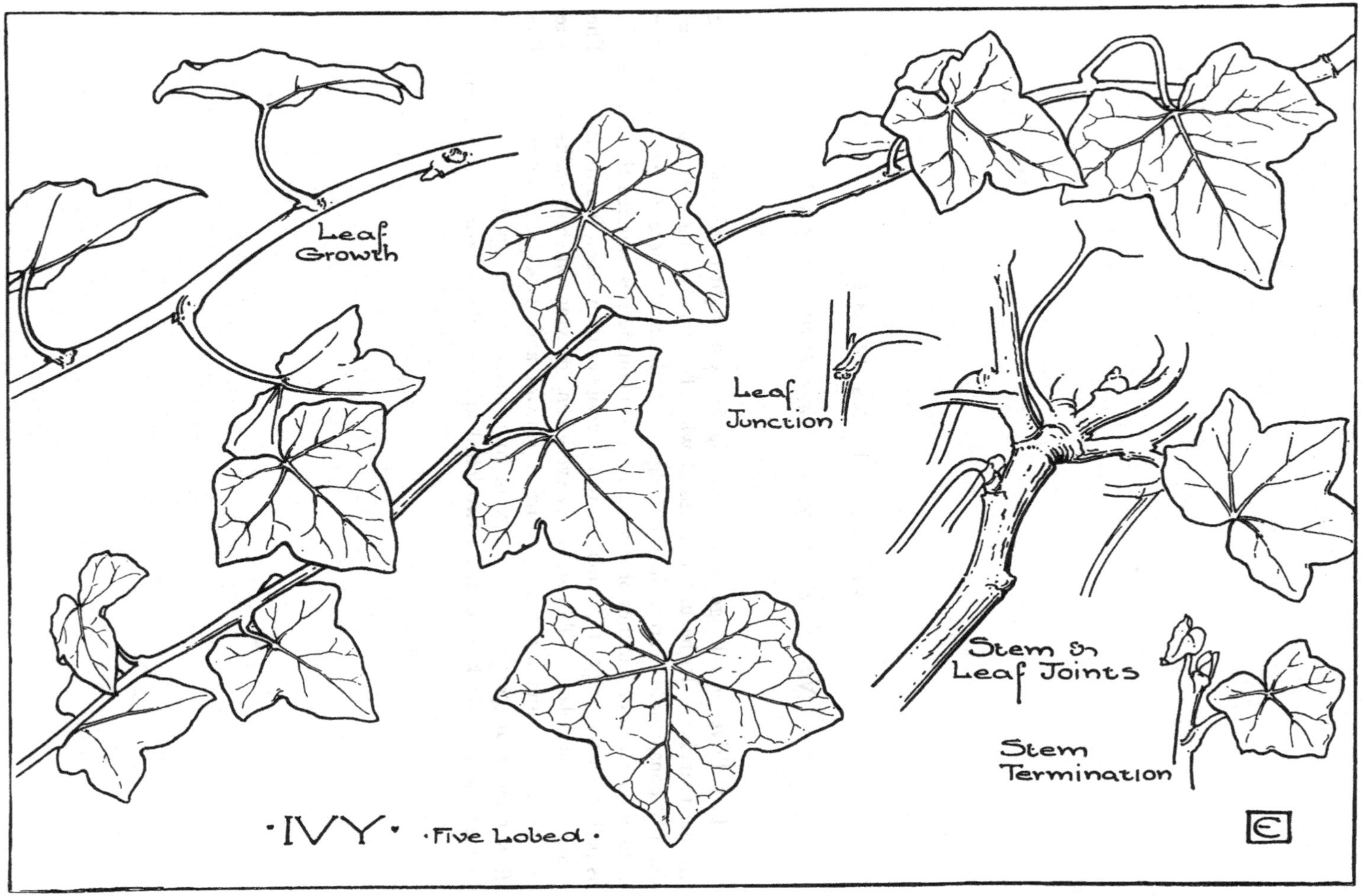

Plate 40.

Leaf
Junction.

Stem &
Leaf Joints

Stem
Termination

Leaf
Growth

·IVY· ·Five Lobed·

PRIVET. Plates 41, 42.

(*Nat. Ord. Oleaceæ. Ligustrum vulgare.*)

THE Privet is a hedge shrub, growing to six or eight feet in height, with long slender branches. Its beautiful groups of white flowers growing in compact panicles form a conspicuous feature in the hedges and thickets where it grows. In the autumn and winter bunches of dry purple succulent berries succeed the flowers. It is perhaps at this stage of greater use to the designer than when it is in flower. The leaves are lanceolate or oblong, nearly evergreen, growing on short stalks in opposite pairs (*decussate*), and are entire. Surface smooth and dark green in colour. The flower has four petals and a calyx of four small teeth, cup-shape. It has two stamens, a short thick style, and a two-lobed stigma. Petals terminate in a tube.

Plate 41.

·PRIVET·

·Leaf·
Form·

Plate 42.

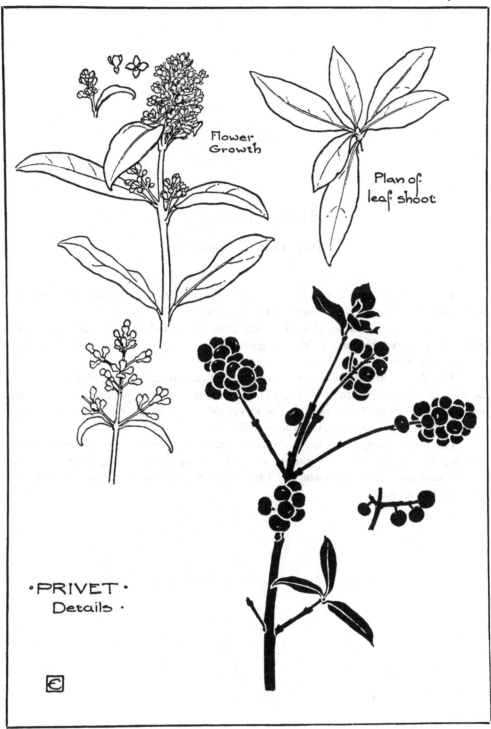

Flower
Growth

Plan of.
leaf shoot

•PRIVET•
Details ·

DAHLIA.   Plates 43, 44.

(*Nat. Ord. Compositæ.*)

THE example here given is one or the single varieties, which offer greater facilities to the artist than the complex double ones.

It is a vigorous-looking, shrub-like plant, bearing a dark-green foliage and rich, variously coloured blooms, making a good show of colour in any garden during August and September.

The flowers are of the composite order, consisting of an outer ring of petals, six or eight in number, the centre being formed of closely packed yellow florets. The calyx is of green sepals, and there is also an epi-calyx. The plant is much branched, and the flowers are borne singly on long stalks.

The leaves are opposite, the lower ones consisting of three or more segments or leaflets, the upper ones are simple, and all are serrated. The fruit forms a conspicuous and interesting feature. The natural size of the flower is from two to three inches across.

Plate 43.

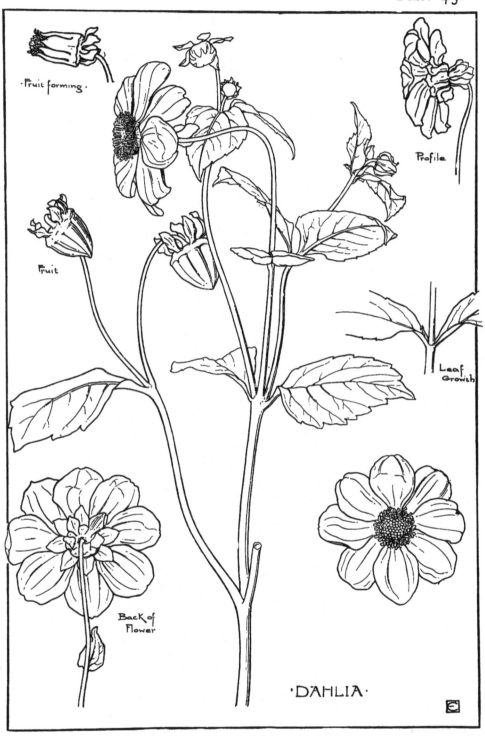

·Fruit forming·

Profile

Fruit

Leaf Growth

Back of Flower

·DAHLIA·

Plate 44.

Leaf in Bud Growth

Leaf in Stalk Junction

Root

· DETAILS OF DAHLIA ·

THE GLADIOLUS.   Plates 45, 46.

(*Nat. Ord. Iridaceæ.   Gladiolus ramosus.*)

THIS plant is usually seen in its cultivated state in our gardens, where it forms a conspicuous feature, assumes a greater variety of colours, and grows to a greater height than it does in its wild state, when it is known as *Gladiolus communis*. The root-stock is bulbous, the leaves those known as linear-lanceolate, and are shorter than the flower-stem. The flower is a perianth, and grows in a one-sided spike; it has six petals or segments which are oblong, three stamens, style three-lobed, larger than the stamens. The flowers are all turned to one side and sessile between two bracts. A spike usually consists of from four to eight blossoms.

Plate 45.

· Spike of
Flower ·

Buds
from the
black

Back
of
Flower

· THE GLADIOLVS ·

Plate 46.

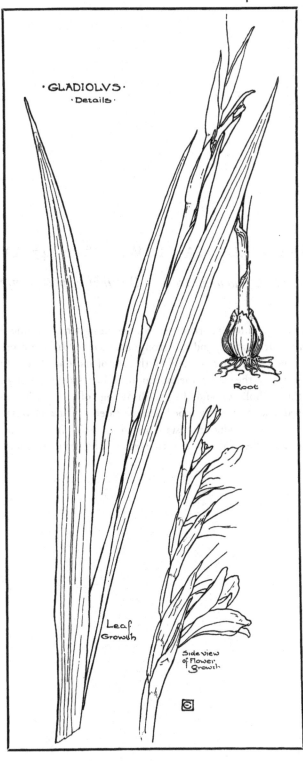

· GLADIOLVS ·
· Details ·

Root

Leaf Growth

Side view of Flower Growth

VEGETABLE MARROW.   Plates 47, 48, 49.

(*Nat. Ord. Cucurbitaceæ.   Cucurbita pepo ovifera.*)

THIS plant offers fine opportunities to the designer; its fine flower and bud forms are extremely decorative, and its strong nervous tendrils and leaves not less so. Creeping along the ground it ties itself by the aid of its tendrils to any other plant in its path, and climbs by the same means up anything strong enough to support it. Its stalks are thick, hollow, and hairy, throwing off, on either side alternately; broad heart-shaped leaves, slightly lobed and serrated. The flowers are yellow, and about three inches across, having the stamens and pistils in separate flowers. The stamens are five, and the anthers are twisted. The style is short and thick, the stigma lobed.

Plate 47.

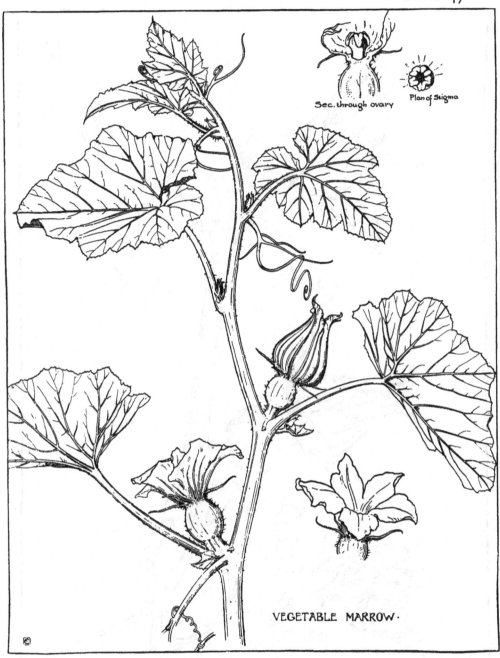

Sec. through ovary   Plan of Stigma

VEGETABLE MARROW.

Plate 48.

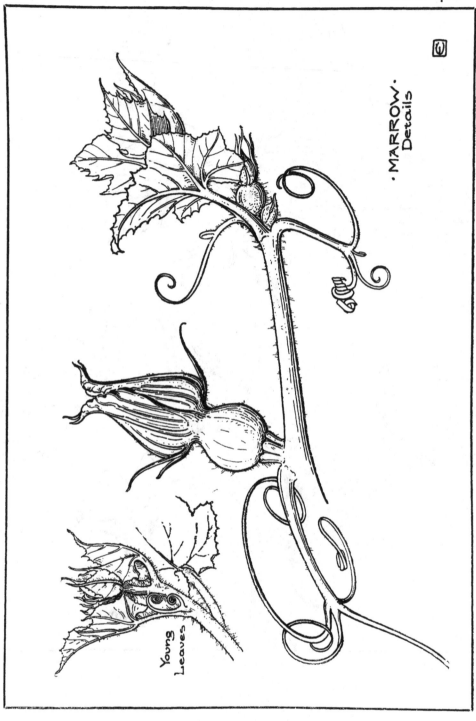

· MARROW ·
Details

Young
Leaves

Plate 49.

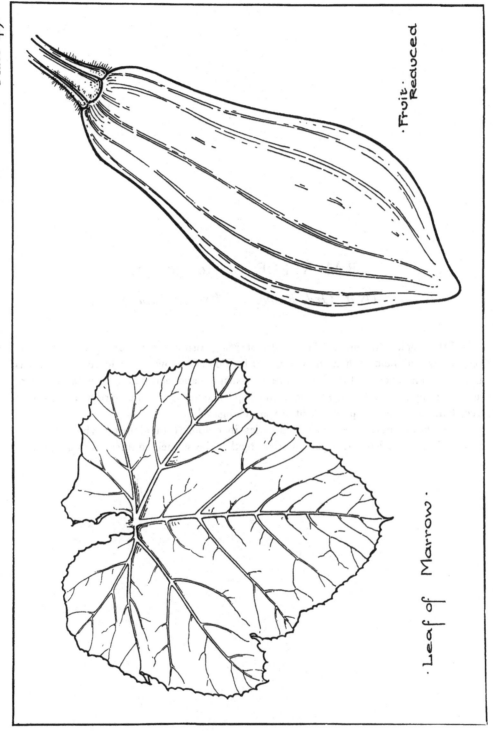

·Fruit·
Reduced

·Leaf of Marrow·

CRAB-APPLE. Plates 50, 51.

(*Nat. Ord. Rosaceæ. Pyrus Malus.*)

THE Apple blossoms in the month of May, forming a picturesque sight in the hedgerows and fields when in bloom. The flowers grow in little tufts or umbels along the branches. The leaves are egg-shaped and serrated. The flowers are succeeded by small fruit, which vary in shape according to the particular variety of tree, some being quite spherical, others are elongated.

The flower bears five petals, and varies in colour from almost a white to a deep pink. The calyx has five sepals or small teeth; the stamens numerous and yellow.

Plate 50.

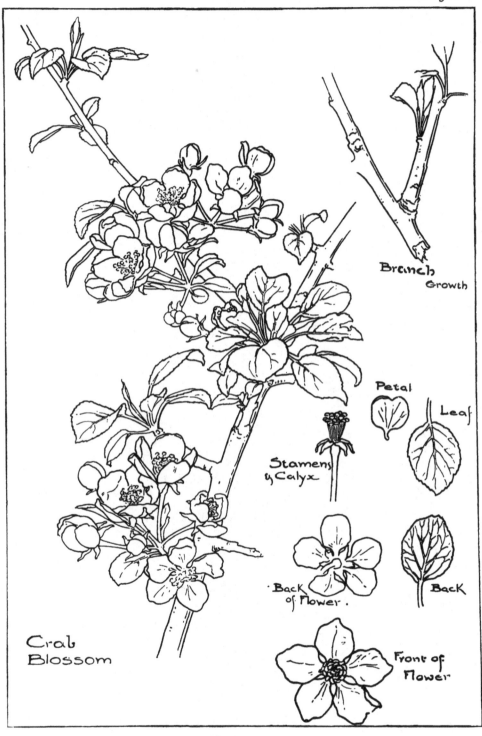

Branch
Growth

Petal

Leaf

Stamens
& Calyx

Back
of Flower.

Back

Crab
Blossom

Front of
Flower

Plate 51.

· Crab Fruit ·

Rounded Fruit

THE CHRYSANTHEMUM. Plates 52, 53, 54.

(*Nat. Ord. Compositæ.*)

THE variety of form and colour to be found in the Chrysanthemum is endless, but the simpler forms are more useful, from the student's point of view, than the more complex ones. The plant is erect and branching, with single flowers on terminal stems, or it is a shrub with flowers in corymbs. The flower-heads are radiating, consisting of a number of long florets, tube-like for a portion of their length, the outer ones being the largest, diminishing in size towards the centre, which is a compact mass of small florets. The calyx is formed of close scaly bracts, and thin pointed bracts occur on and at the base of the flower-stalks. The bud forms are very decorative features, and frequently grow from the axils of the leaves, or are surrounded by a leafy whorl. The leaves vary a good deal on different plants, but all have the same broad characteristics ; they are mostly lobed, some of them almost to the midrib, and deeply serrated, with broad and narrow stipules at their junctions, some of which take the form of separate small leaves, and others are wing-like appendages to the petiole. The growth up the stem is in this case a spiral. The colour of the flower figured here is a pale yellow ; its form will be found sufficiently complex for the student to deal with ; in fact, the nearer it approaches its original primitive form, as in the Ox-eye Daisy, the more manageable it is for decorative work. The student should look carefully at the separate studies of the leaf growth given on the two succeeding plates.

Varieties of the plant may be found from May to December ; the example here given is an autumn-flowering variety.

Plate 52.

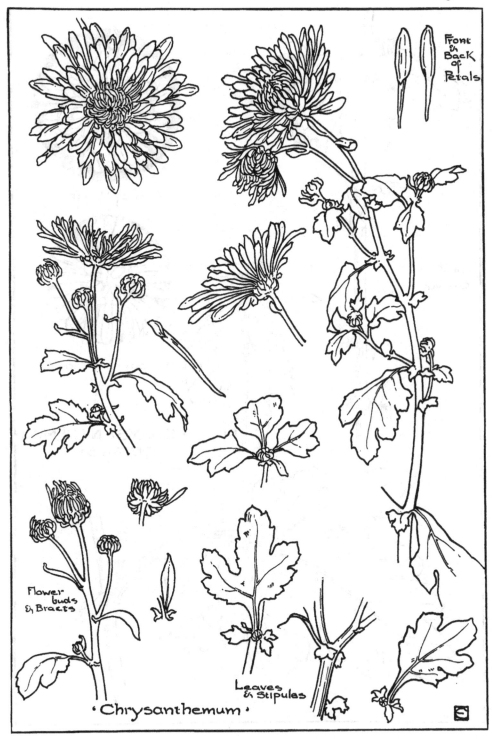

Front
&
Back
of
Petals

Flower
buds
& Bracts

Leaves
& Stipules

'Chrysanthemum'

Plate 53.

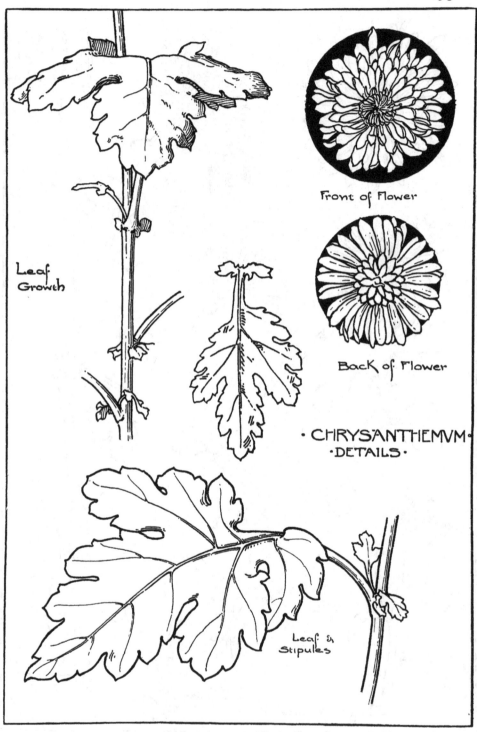

Front of Flower

Back of Flower

Leaf
Growth

·CHRYSANTHEMVM·
·DETAILS·

Leaf &
Stipules

Plate 54.

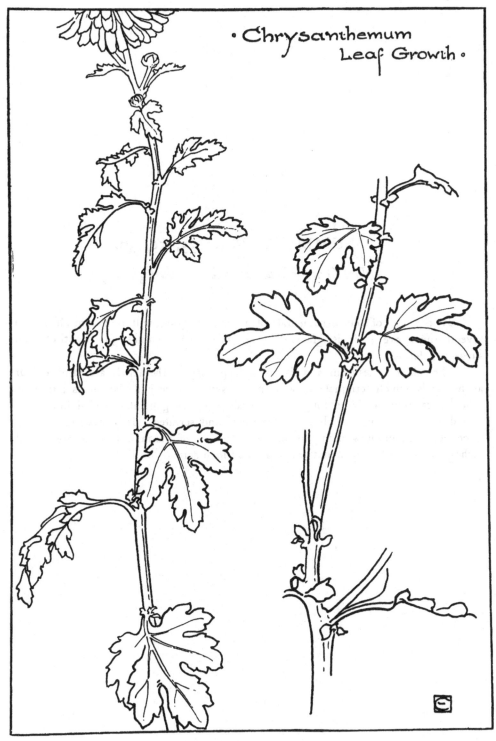

Chrysanthemum
Leaf Growth

COLUMBINE. Plates 55, 56, 57.

(*Nat. Ord. Ranunculaceæ. Aquilegia.*)

By a study of this ornamental plant, the designer will find himself repaid several times over ; it is graceful in growth, and abounds in beautiful and suggestive forms.

The colour varies from a white to a deep purple ; the sepals are same colour as the petals, which terminate in a horn-like spur ; these and the sepals alternate The leaves grow chiefly on long radical stalks, and are generally divided into three very distinct segments, which are again lobed and serrated. It grows from one to three feet high, and flowers in June and July. Some of the radical leaves are several inches across, the flowers about one and a half inch across.

Plate 55.

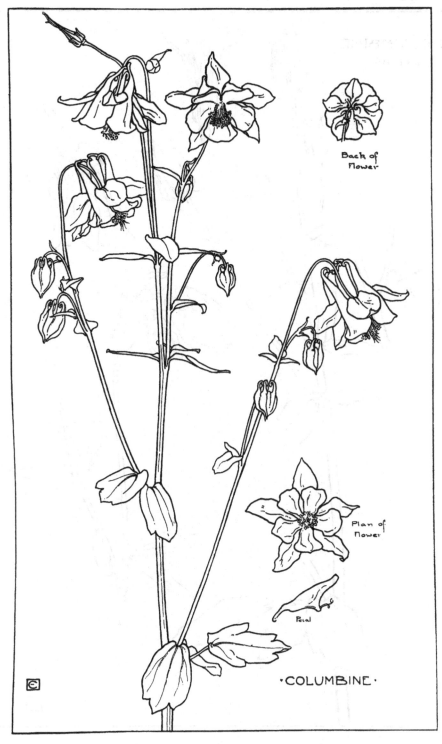

Back of
Flower

Plan of
Flower

Petal

·COLUMBINE·

Plate 56.

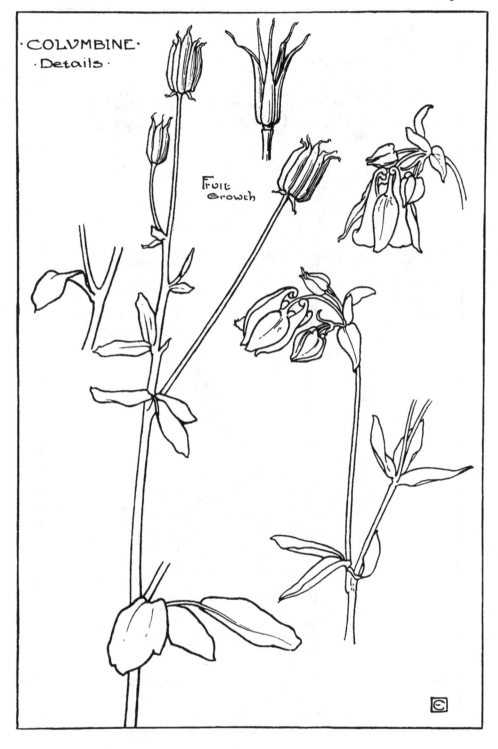

·COLVMBINE·
·Details·

Fruit
Growth

Plate 57.

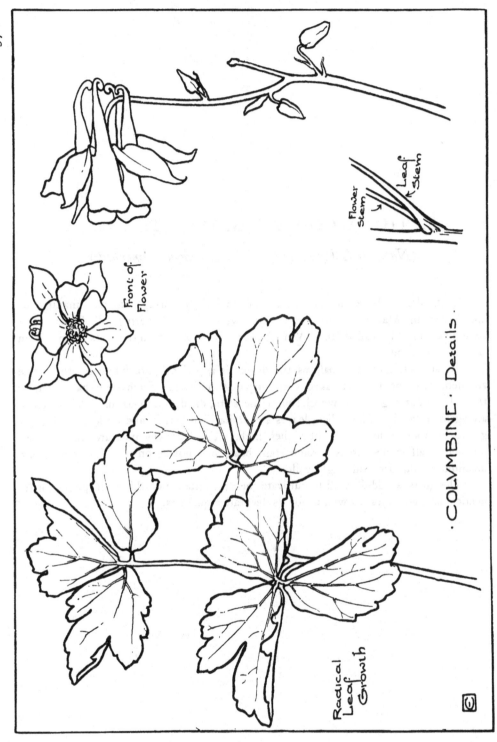

Front of
Flower

Flower
Stem

Leaf
Stem

Radical
Leaf
Growth

· COLVMBINE · Details ·

H

THE WOOD ANEMONE. Plate 58.

(*Nat. Ord. Ranunculaceæ. Anemone nemorosa.*)

THIS dainty little flower is one of the early spring blooms, making its appearance in March in the copses and woods, and flowering till June. It is sometimes prettily called the Wind-flower. In fact, it derives its name from *Anemos*—the wind.

Its six petals, or its sepals, as the flower is a perianth, are white on the inside and delicately tinted and streaked with purple at the back. Each stem has generally three leaves, arranged in a whorl round the stem, divided into as many lobes, which are again serrated. The radical leaves are similar, but larger and on long stalks. It grows from about four to eight inches high. The blossoms measure from one to one and a half inches across. Each stem bears one flower only. The stamens are indefinite in number, and bright yellow.

Designers would do well to cultivate the acquaintance of this little plant, for its poetical suggestiveness as well as for its decorative qualities.

Plate 58.

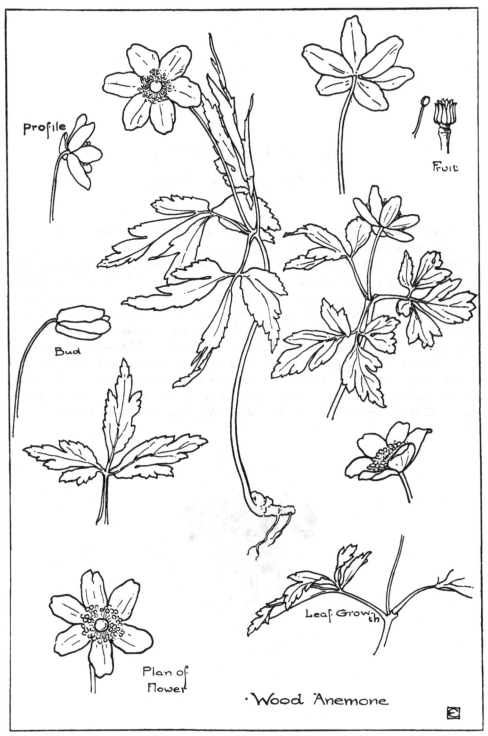

Profile

Fruit

Bud

Plan of
Flower

Leaf Growth

·Wood Anemone

THE VIOLET. Plate 59.

(*Nat. Ord. Violaceæ. Viola.*)

THIS is one of the several wild varieties of the Violet, one or other of which colours the hedgebanks and woods from March to July. From the designer's point of view it matters not whether it be the wood-violet, the sweet-violet, the dog-violet, or the hairy-violet, as they are all equally beautiful in form. The flowers and leaves grow practically straight from the root on several varieties, but on the one figured here, they grow from the stems also. The usual height is from about three to five inches. The flowers have five petals and five sepals, the centre petal being the largest and ending in a spur at the back of the flower. The colour varies from a pale blue to a deep purple. The leaves are broad and heart-shaped or cordate, and serrated, rather dark green in colour. There are five stamens, the anthers of which and the pistil are united. The stipules are long and pointed. The fruit is a capsule of three divisions. Found on hedgebanks, in thickets, and pastures.

Plate 59.

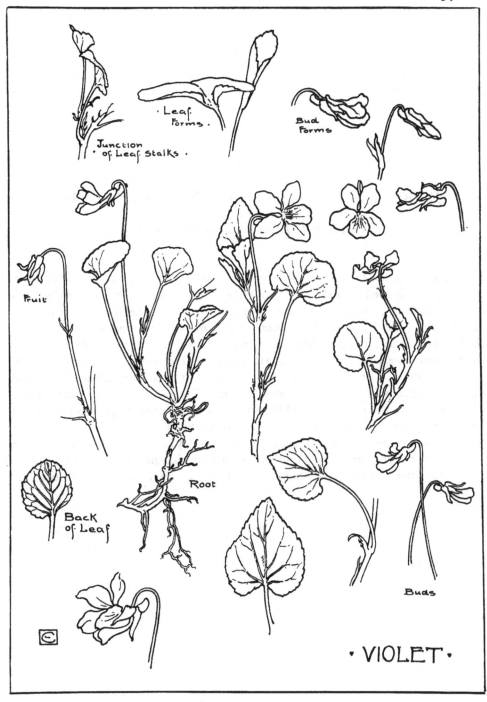

Leaf Forms.

Junction of Leaf Stalks.

Bud Forms

Fruit

Back of Leaf

Root

Buds

· VIOLET ·

THE BLUEBELL, or WILD HYACINTH.   Plate 60.

(*Nat. Ord. Liliaceæ. Scilla nutans.*)

THIS must not be confounded with the Scottish Bluebell, which is the same as the English Harebell. In the early spring the leaves of the Hyacinth begin to break through the earth, and are soon followed by the pale buds of the flower. Its root is a small bulb. The leaves are linear and slightly grooved, growing straight from the bulb and attaining in cases to fully a foot in height, but they do not reach the height of the flower-stalk. The growth of the flowers is a terminal, one-sided raceme of drooping flowers. The flower itself is a perianth, and is a long bell-shape, divided into six segments, reflexed at their extremities. There are six anthers attached to the inside of the flower, a style, and stigma. The bracts grow in pairs at the base of the flower stalks. The fruit is a capsule containing black shiny seeds. This flower is to be found in plenty in woods and copses during May and June, giving a rich depth of colour to the undergrowth, and furnishing, one may imagine, a practical illustration to the fairy tale that the " sky has fallen." It is a graceful and useful plant to the designer.

Plate 60.

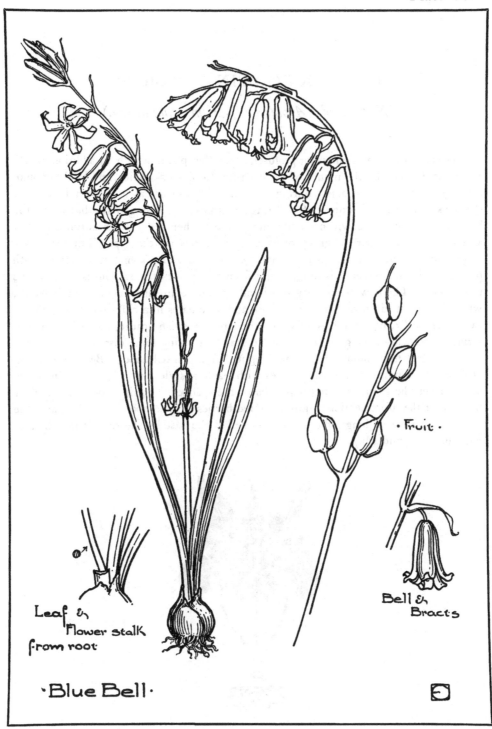

·Fruit·

Bell &
Bracts

Leaf &
Flower stalk
from root

·Blue Bell·

THE CUCKOO-PINT. Plate 61.

(*Nat. Ord. Aroideæ. Arum maculatum.*)

DESPITE the commonness or profuseness of this plant, which in addition to the above name, bears also the names of Wild Arum, Lords-and-Ladies, and Wake-Robin, it is not a very common object during its term of flowering, as it is then hidden by the thick covering and growth of the hedge bottoms where it grows; but when the hedges have thinned down in the autumn you may then see it in its autumn garb— a short thick stem with a group of bright red berries on the top. The leaves are large and glossy, and arrow-head shape or *ovate-hastate*, sometimes spotted with purple. The inflorescence is insignificant from our point of view, but is nevertheless interesting. The flowers are very small and are borne in rings on a stout column, which is continued upward to some length and terminates in a club-like head; this is called the *spadix*, near the base of which the large greenish spathe, like a large bract, springs and encloses the greater part of the spadix, growing some distance beyond the top of it, but not completely surrounding it, like a pointed hood. Beneath the ring of florets are a number of sessile ovaries with each a style and stigma. When the berries ripen the spathe and upper portion of the spadix have fallen away, leaving for the winter the aforementioned group of red berries. The root is a tuber, and the leaves are carried on long radical stalks clothed at the base by sheaths. It may be met with in April and May.

Plate 61.

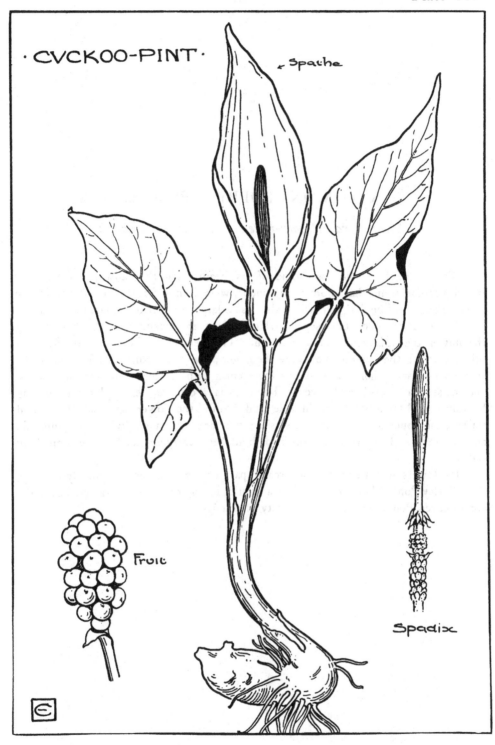

·CVCKOO-PINT·

Spathe

Fruit

Spadix

COMMON POPPY.   Plate 62.

(*Nat. Ord. Papaveraceæ.   Papaver Rhæas.*)

THE cornfield would lose some of its beauty, and the railway embankment would become somewhat more monotonous, were it not for the abundance of the scarlet Poppy.   The designer would also lose a large amount of valuable material.

It has a calyx of two sepals, which fall at the opening of the bud, giving the flower the appearance of a perianth.   The petals are four, and usually of a bright scarlet; the two outer ones are the larger; the buds hang gracefully down the stem covered with short hairs, as are the stems.   The stamens are very numerous, and the stigma is sessile and raved.   The fruit is a capsule, nearly globular, opening by valves.   The leaves vary in size and form; the lower ones are large and stalked, the upper ones stalkless or sessile; they are very much divided or pinnatifid and serrated.   The plant is erect and branched, carrying its blossoms on long stalks.

Its flowers may be found the greater part of the summer, but mostly in July. The Shirley Poppy (*Papaver Rhæas*) is a hybrid, having the same characteristics as the common red Poppy, but a greater variety of colours.

Plate 62.

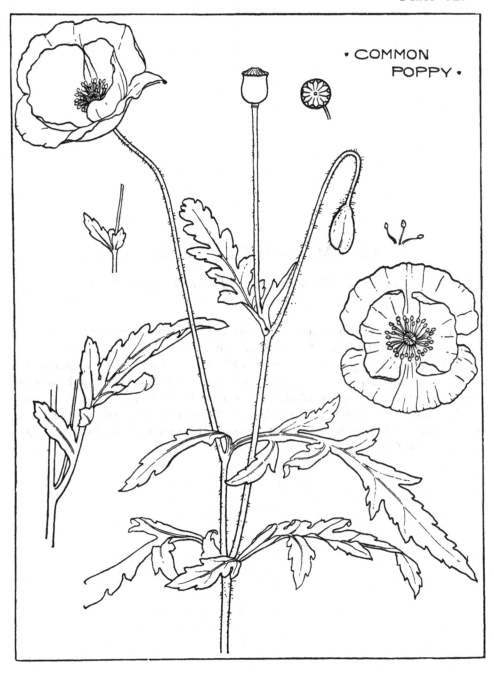

• COMMON
POPPY •

OPIUM POPPY. Plates 63, 64.

(Nat. Ord. Papaveraceæ. Papaver somniferum.)

This is usually an occupant of the garden; occasionally one is found growing in a wild state, but this is mainly due to accident.

The foliage is very beautiful, but probably not so useful to the designer as that of the common Poppy; it is of a glaucous green, and the leaves clasp the stem by their base; they are slightly lobed, and are irregularly toothed. The flowers are usually bluish-white, with a purple splash at the base of the petal, but the colour varies a deal. The capsule is large and globular. Flowers July and August. Sepals two and petals four.

Plate 63.

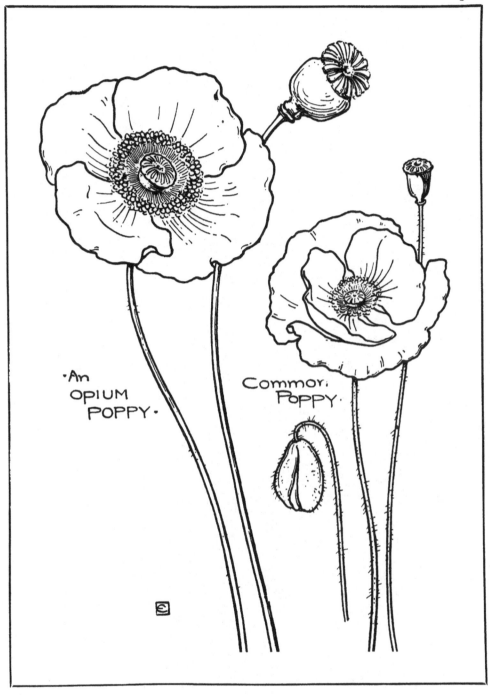

·An
OPIUM
POPPY·

Common
POPPY·

Plate 64.

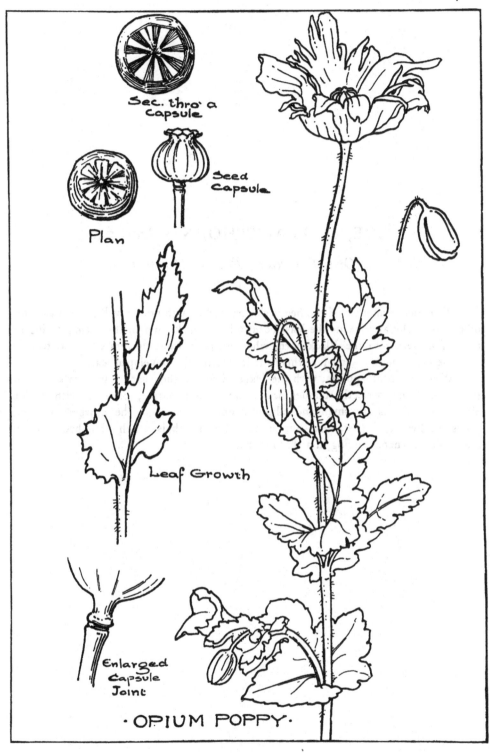

Sec. thro' a
Capsule

Plan

Seed
Capsule

Leaf Growth

Enlarged
Capsule
Joint

· OPIUM POPPY ·

SLOE, or BLACKTHORN. Plate 65.

(*Ord. Rosaceæ. Prunus spinosa.*)

THE Sloe is another of the hedge shrubs, and though not so well known as some others, is not by any means the least beautiful. It is known in the spring during its term of flowering as the Blackthorn, when its pure white flowers contrast strongly with the black thorny stems; at this time it is without a developed leaf.

When the fruit appears, with the leaves, later in the year, it is then known as the Sloe. The perfectly rounded fruits are a deep purple, covered with a rich bloom. Its leaves are small and elliptical, finely serrated. The flowers have five petals and five sepals; stamens numerous. The shrub is much branched, some of the smaller branches ending in a sharp thorn.

Plate 65.

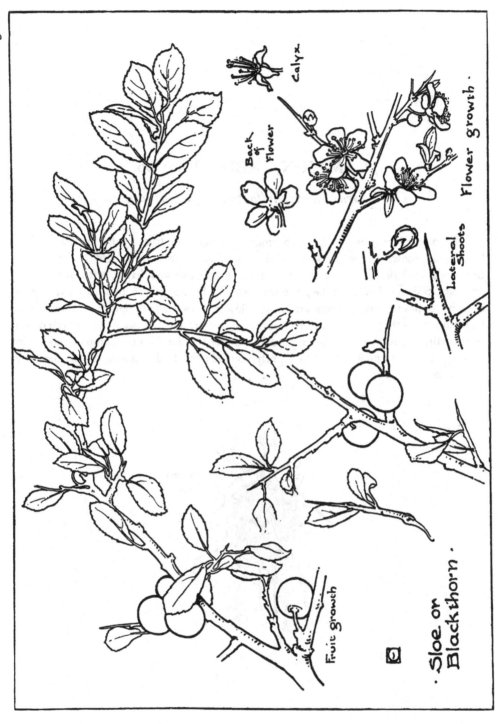

Calyx

Back of Flower

Flower growth.

Lateral Shoots

Fruit growth

· Sloe or Blackthorn.

I

CAMPION, RED.   Plate 66.

(*Nat. Ord. Caryophyllaceæ.   Lychnis diurna.*)

Found plentifully in hedge-bottoms and copses from June to September. Commonly known as the Cuckoo.   Colour of flowers is from a light to deep pink; five petals, each of which is cleft; the calyx is purplish and hairy, with five angular divisions, connected into a tube; stamens ten.   The leaves are pointedly oval, grow in pairs opposite, upper ones are sessile, lower ones stalked, entire.   The White Campion is another variety, the flowers being larger, and the plant generally coarser in growth.   The calyx is also larger and rounder.   The White Campion opens its blossoms in the evening air, and is slightly scented; the Red blossoms all the day, and is scentless.

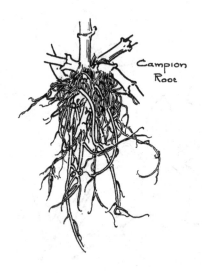

Campion
Root

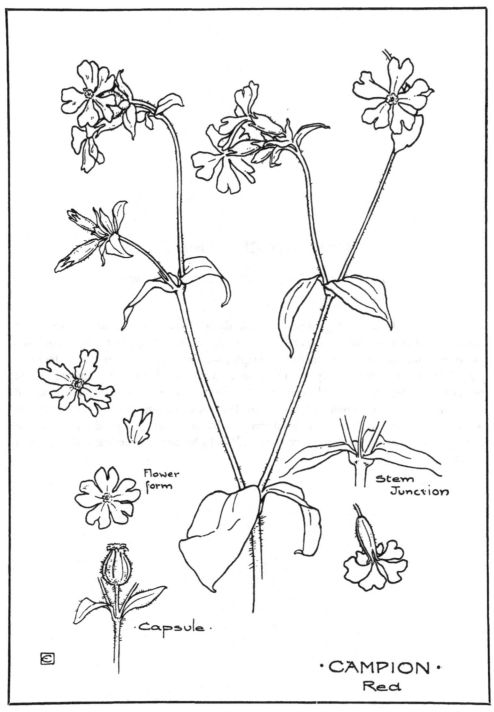

Plate 66.

Flower
form

Stem
Junction

·Capsule·

·CAMPION·
Red

GROUNDSEL. Plate 67.

(*Nat. Ord. Compositæ. Senecio vulgaris.*)

THIS is a little plant growing abundantly in Britain as a common weed, but singularly attractive from a decorative point of view. It is erect and branching, growing from six to twelve inches in height, with pinnatifid leaves having toothed, jagged lobes, growing spirally up the stem; some of them being stalked, while others are sessile, clinging to the stem by their bases. The flower-heads are terminal corymbs or clusters of small yellow-headed flowers, consisting only of tubular florets without any outer ray or petal whatever. The involucre is cylindrical, of about twenty bracts, having smaller outer ones. It is to be found practically all the year round.

Plate 67.

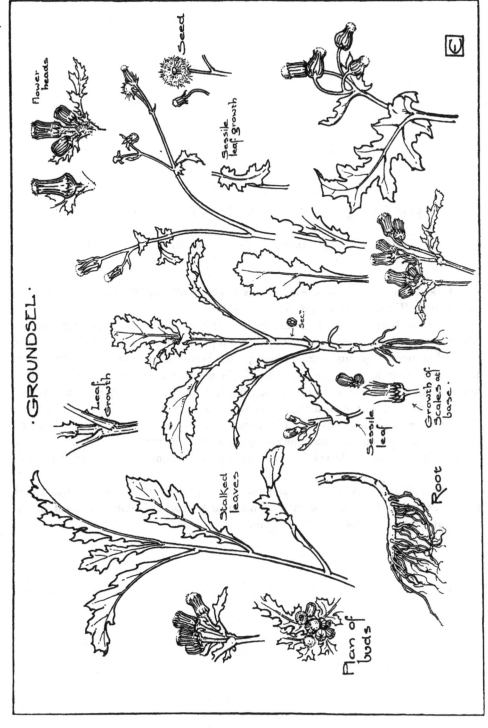

THE AUTUMN ANEMONE.   Plates 68, 69.

(*Nat. Ord. Ranunculaceæ.   Anemone japonica.*)

THE Autumn or Japanese Anemone is in nearly all respects, except that of size, like the little Wood Anemone (*Anemone nemorosa*).

It grows from one to two feet high in rather a loosely branching manner. When in blossom its graceful growth and white, as in this case, or pinkish blooms, and its dull green vigorous-looking leaves, offer beautiful material to the student of design and plant form.   The flower itself is a perianth somewhat irregular in form, consisting of six or more petals or sepals, delicately tinged with purple on the outside. The anthers are numerous and bright yellow, forming a ring of gold round the green globular stigma.   The leaves are radical, excepting three involucral leaves placed on the stem at the junction of the flower-stalks, which are from six to eight inches in length.   This whorl of leaves or bracts is one of the characteristics of the whole genus.

The radical leaves are large and three-lobed, rather sharply serrated, and with pronounced veining.   The buds are roundish and vigorous-looking, enclosed by the three outer sepals.   Flowers in the autumn.

Plate 68.

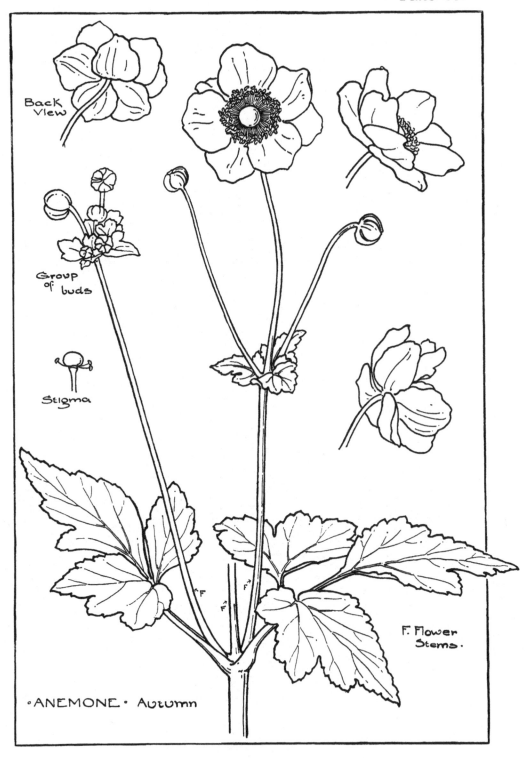

Back View

Group of buds

Stigma

·ANEMONE· Autumn

F. Flower Stems.

Plate 69.

Leaf &
Stalk
Junction

·ANEMONE· Details·

·Radical
Leaves·↗

WILD TEASEL. Plates 70, 71, 72.

(*Nat. Ord. Dipsacaceæ. Dipsacus sylvestris.*)

THE heads have numerous slender spiny bracts, from eight to twelve forming an involucre at base ; they are curved upwards, ending in a sharp point. The flowers are purple and expand in irregular patches on the head ; the corolla is tubular and consists of four unequal lobes. The stem leaves are opposite and not stalked ; the lower couples are joined together by their bases, *i.e. connate*, forming a cup in which the water collects. They grow from three to six feet high. The stems are angular and spiny. It is the general growth and form of this plant, the beautiful heads and bracts, which make it a successful designer's plant. To be found on the roadsides and waste places. Flowers in the late summer and autumn.

Plate 70.

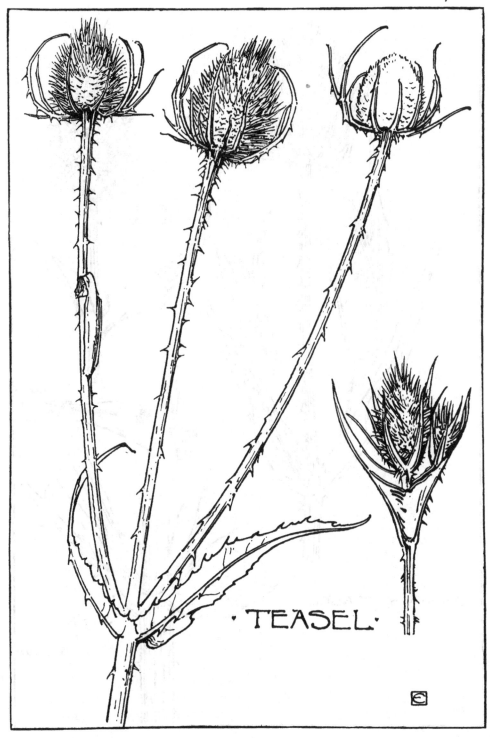

· TEASEL ·

Plate 71.

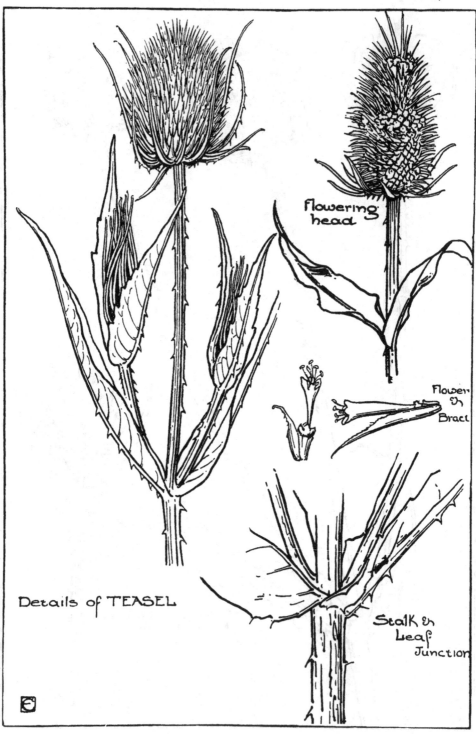

flowering
head

Flower
&
Bract

Details of TEASEL

Stalk &
Leaf
Junction

Plate 72.

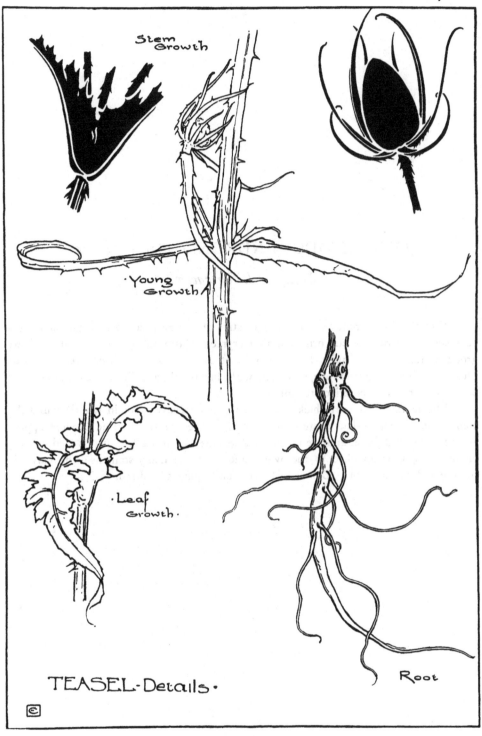

Stem Growth

Young Growth

Leaf Growth

Root

TEASEL·Details·

THE RHODODENDRON. Plates 73, 74.

(*Nat. Ord. Ericaceæ.*)

THE Rhododendron, although not a native of Britain, has formed for so long a conspicuous feature as an ornamental evergreen shrub in our parks and private grounds, that we can quite look upon it, at least, as a naturalised tree. In our illustrations of it only the foliage is dealt with, but which, as will be seen, offers great possibilities to the decorative student.

The leaves, which are thick and glossy, grow spiral-wise or in whorls round the stems, and are oblong-lanceolate in form, with entire margins. The flowers appear early in May and June, and are in terminal groups ; each has five petals or divisions, ending in a tube attached to the ovary, and twice as many stamens ; the calyx is minute and five-toothed. Colour of flowers pink, purple, and red.

Plate 73.

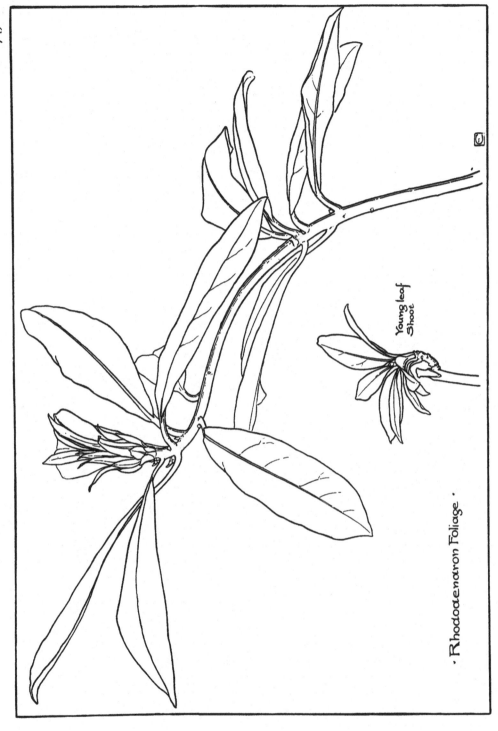

Young leaf
Shoot

· Rhododendron Foliage ·

Plate 74.

Plan

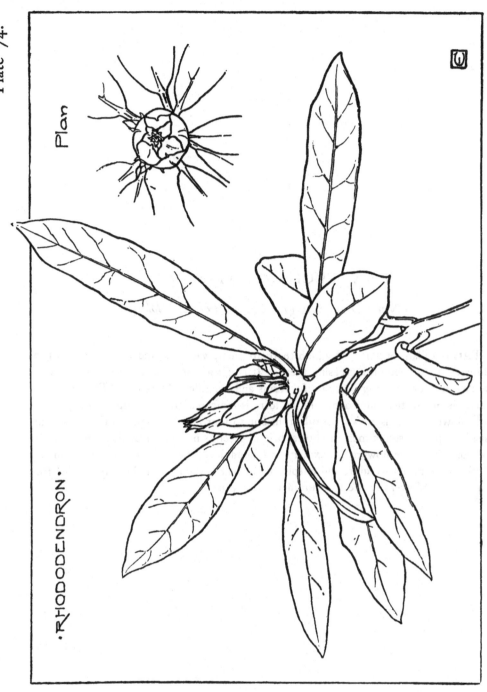

·RHODODENDRON·

K

THE FUCHSIA. Plates 75, 76.

(*Nat. Ord. Onagraceæ. Fuchsia globosa.*)

THIS is a plant or shrub of which there are many varieties, and the tendency of the gardeners is to keep on increasing them.   It is not a plant that has been overdone by the designer, considering its beautiful forms and pendulous blossoms.   This particular variety is one of the older ones, and its form is indicated by its Latin name, *globosa*— globe-flowered.   It is a dwarf shrub from one to four feet in height.   The leaves are usually opposite, sometimes in threes, bright green, smooth, serrated, ovate in form, and pointed.   The flowers are axillary and pendulous, the calyx is four-cleft and rich crimson, the petals are four, and purple, about half the length of the calyx.   The fruit is a berry.   Flowers in the late summer and autumn.

Plate 75.

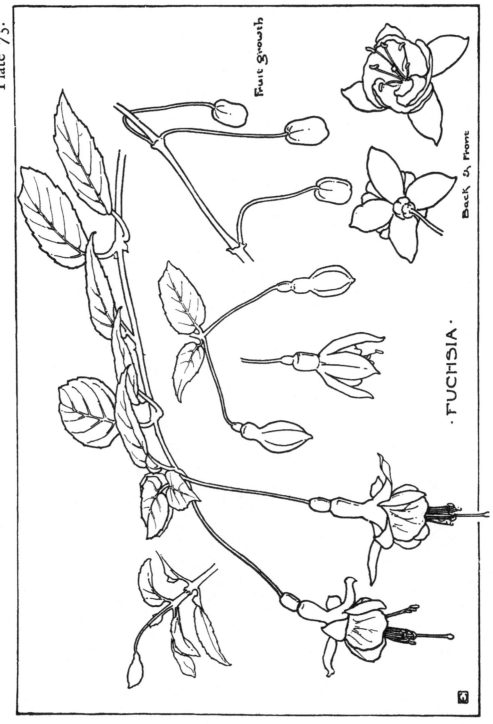

Fruit growth

Back & Front

· FUCHSIA ·

Plate 76.

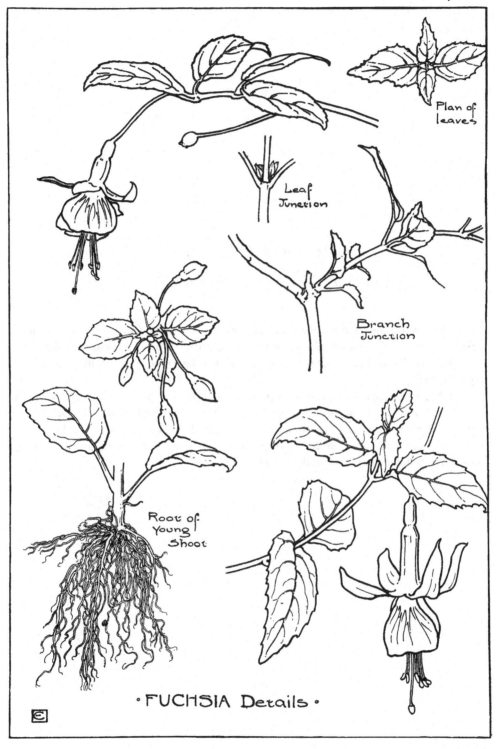

Plan of leaves

Leaf Junction

Branch Junction

Root of Young Shoot

• FUCHSIA Details •

WOOD SORREL. Plate 77.

(*Nat. Ord. Geraniaceæ. Oxalis corniculata.*)

THIS is of the yellow flower-bearing species, carrying from two to five pale yellow flowers on one slender stem, forming an umbel. The flowers are smaller than the common Wood-sorrel (*O. Acetosella*), which is to be found in moist shady woods in April and May.

It has five sepals, five petals, ten stamens, and stigmas five. The stalk is much branched. Flowers from June to September. The other yellow variety (*O. stricta*) has a more erect stem, and has from two to eight flowers on one peduncle. The leaves of the Oxalis are trefoils, consisting of three obcordate yellow-green leaflets, which droop down on to the stalk at night.

Plate 77.

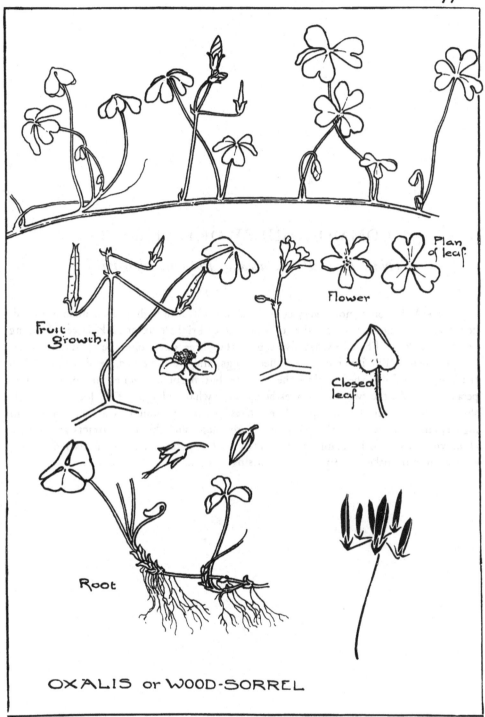

Plan of leaf

Flower

Fruit growth.

Closed leaf

Root

OXALIS or WOOD-SORREL

COMMON MILKWORT. Plate 78.

(*Nat. Ord. Polygaleæ. Polygala vulgaris.*)

THE Milkwort is not a very conspicuous plant, although it is abundant in this country. Its inflorescence consists of a raceme of bright blue or pink flowers hanging on short stalks. The calyx has five sepals, the three outer ones being small, linear, and greenish. The inner two are called wings, and are large and petal-like. When the flower has fallen they enclose the capsule, but become greener in colour. The petals are small, the two lateral ones being somewhat oblong. The lowest is keel-shaped, and crested at the tip. The leaves are more numerous at the base than higher, and also broader ; those higher up are longer and thinner. Bracts at the base of flower stalks. The stamens are united to the petals, and are in two divisions, each with four anthers. Style one, with single stigma.

Plate 78.

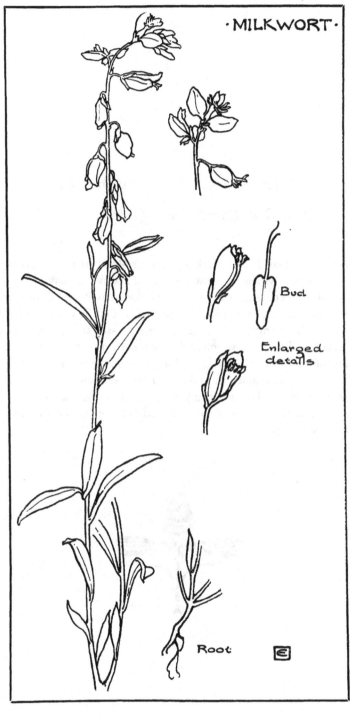

·MILKWORT·

Bud

Enlarged details

Root

THE OAK. Plates 79 and 80.

(*Nat. Ord. Cupuliferæ.   Quercus Robur.*)

SPEAKING of the Oak is like speaking of one's oldest friends. Perhaps no school-boy remembers the time when he knew nothing of acorns and oak-apples.

But to speak of them from a decorative standpoint will appeal to a much smaller but not a less appreciative community.   The British Oak is the longest-lived among the native trees of our islands.   The leaves are short-stalked, sometimes sessile, and are obtusely lobed.   The fruit, which is the well-known acorn, is carried in a shallow cup, consisting of a series of closely-knitted scales forming an involucre.   They grow sometimes in clusters, sometimes in spikes.   The tree flowers in the spring, when the leaves are growing.   There are many varieties, and the leaves vary considerably. The Oak was frequently used as an ornamental motive by the ancients, and by the Artists of the Middle Ages.

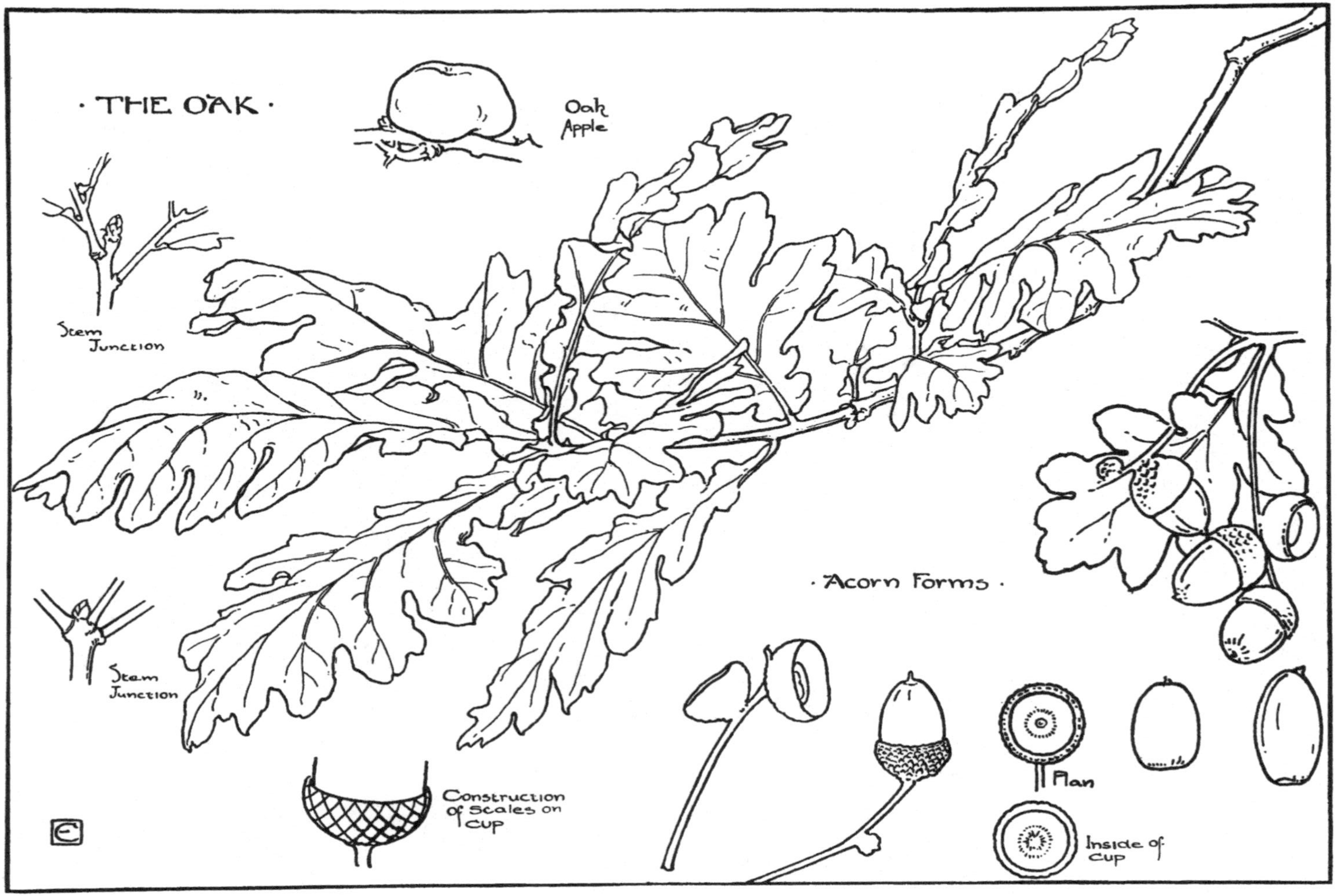

Plate 79.

. THE OAK .

Oak Apple

Stem Junction

Stem Junction

. Acorn Forms .

Plan

Inside of Cup

Construction of Scales on Cup

Plate 80.

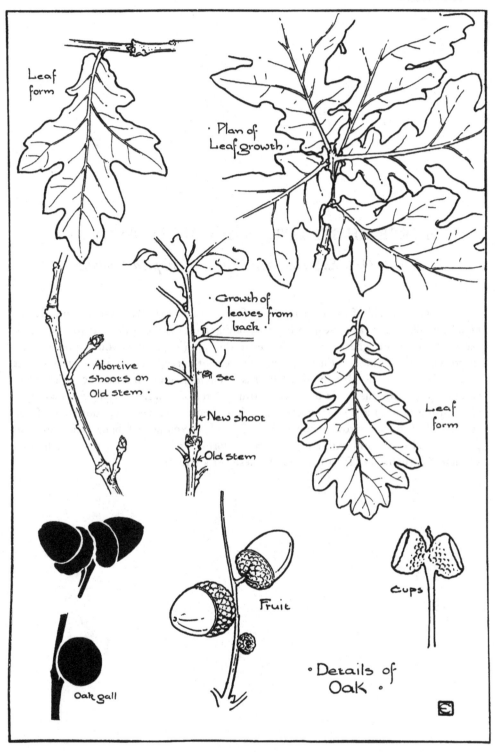

Leaf form

Plan of Leaf growth .

Growth of leaves from back .

Abortive Shoots on Old stem .

Sec

New shoot

Old stem

Leaf form

Oak gall

Fruit

Cups

Details of Oak .

HORSE-CHESTNUT.   Plates 81, 82, 83, and 84

(*Nat. Ord. Sapindaceæ.   Æsculus hippocastanum.*)

WHAT was said of the Oak might as fitly be said of the Horse-chestnut. It has figured much oftener as a weapon in schoolboy games than as a decorative feature in the hands of the designer.   But its power in the latter capacity is much greater than its power in the former. The schoolboy would not agree, but the artist knows ! It is not a native of the British isles, but has been long enough introduced (1629) to become quite naturalised. It bears a beautiful cluster of flowers, varying from white to pink, in the form of a panicle or thyrsus.   The leaves are palmate, consisting of five or seven leaflets, and grow in opposite pairs.   The young leaf buds (illustrated on plate 84) and fruit form extremely decorative forms, and the flower, which is figured on plate 83, should not fail to attract the designer's attention.

Plate 81.

Bursting
Fruit

Opening Fruit ·

· HORSE CHESTNUT ·

Plate 82.

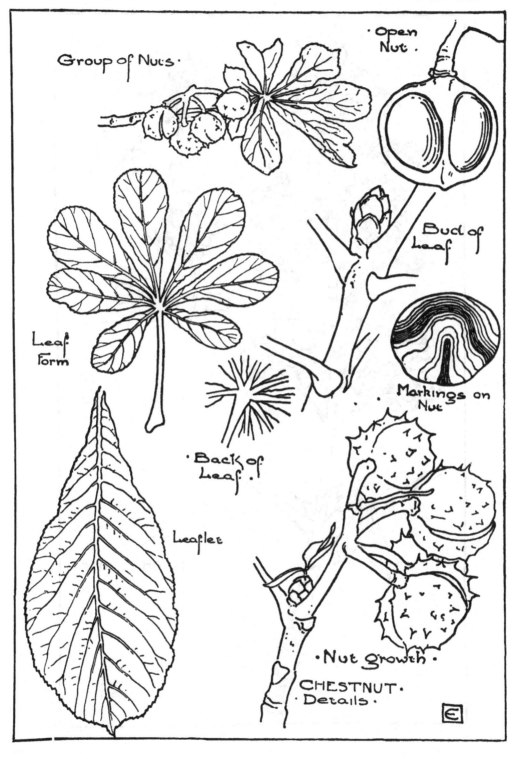

Group of Nuts.

· Open Nut ·

Bud of Leaf

Leaf Form

Markings on Nut

· Back of Leaf ·

Leaflet

· Nut growth ·

CHESTNUT · Details ·

Plate 83.

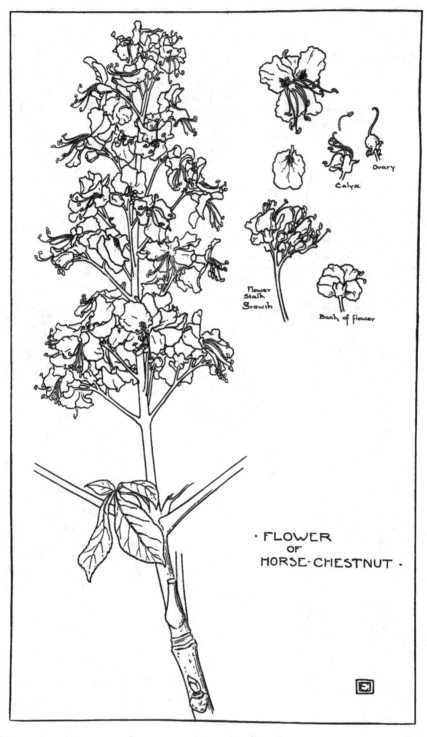

Calyx

Ovary

Flower
Stalk
Growth

Back of flower

· FLOWER
OF
HORSE-CHESTNUT ·

Plate 84.

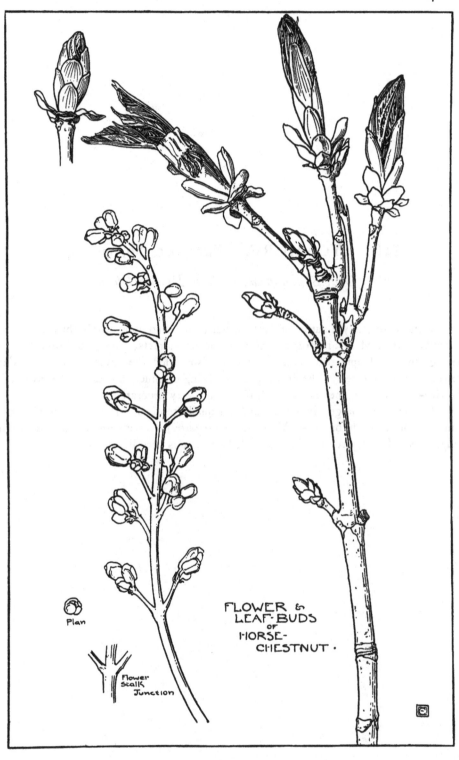

Plan

Flower
Stalk
Junction

FLOWER &
LEAF-BUDS
OF
HORSE-
CHESTNUT.

THE SYCAMORE.   Plates 85, 86, and 87.

(*Nat. Ord. Sapindaceæ.   Acer Pseudo-Platanus.*)

THIS is not really a native of Britain, but it has been long established here.   It is extensively used as an ornamental tree for our streets, parks, and squares.   It belongs to the Maple tribe (*Acer*), and is also called the greater Maple.   The flower is only indicated in the drawing as not being of much use to the designer, but the two-winged fruit, the leaves and leaf buds are very decorative.

The leaves are large, five-lobed, and unequally toothed.   It flowers in May and June.   The English, or lesser Maple (*Acer campestre*) is very common, and is in our hedgerows.   It is figured on Plate 85.   The foliage is very beautiful.

Plate 85.

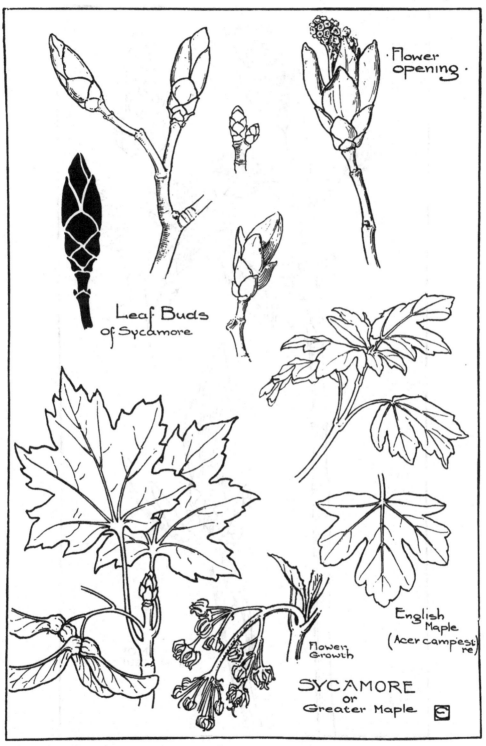

Flower opening ·

Leaf Buds
of Sycamore

English Maple
(Acer campestre)

Flower Growth

SYCAMORE
or
Greater Maple

Plate 86.

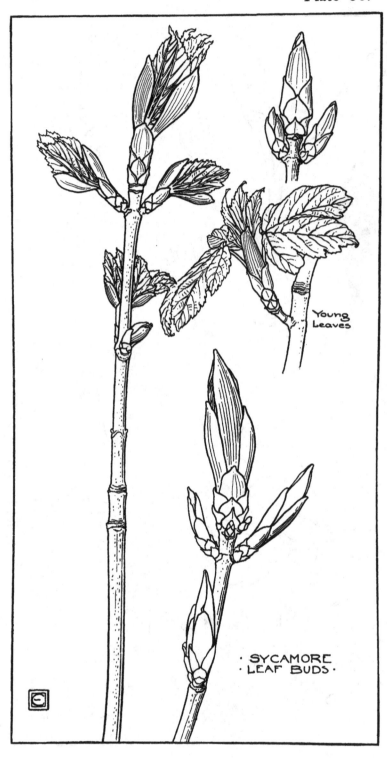

Young
Leaves

· SYCAMORE ·
· LEAF BUDS ·

Plate 87.

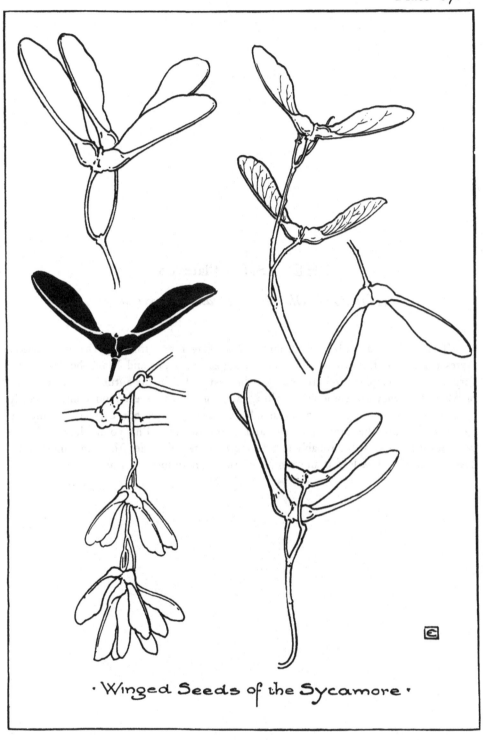

· Winged Seeds of the Sycamore ·

THE ASH. Plate 88.

(*Nat. Ord. Oleaceæ. Fraxinus excelsior.*)

THE Ash is a native of Britain. The Privet and this are the only native representatives of this order. The flowers appear in April and May, the leaves are very late in developing, and are amongst the first to fall in the autumn. The fruits, or " keys " as they are sometimes called, each contain two seeds, and consist mainly of a membranous wing. The flowers have neither calyx nor corolla, but merely consist of the essential organs, a pistil and two stamens. The young leaf buds are very beautiful, and will probably appeal more to the designer than the more fully developed leaf. The leaf is an example of the pinnate form of leaf.

Plate 88.

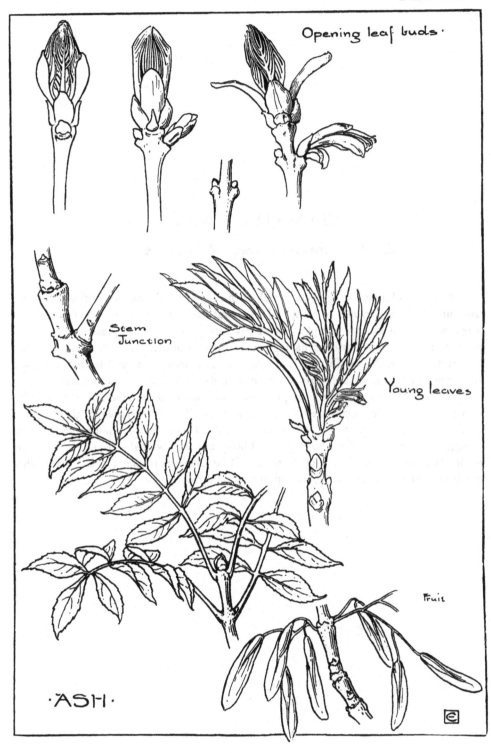

Opening leaf buds.

Stem
Junction

Young leaves

Fruit

·ASH·

CLEMATIS. Plate 89.

(*Nat. Ord. Ranunculaceæ. Clematis viticella.*)

THIS is a climber which makes rich with its purple blooms any trellis or house-front up which it is trained to climb. It is useful from the designer's point of view in many ways; it is not too complex, as are most cultivated flowers, possessing but four petals or sepals—some varieties having six or eight—and is a perianth. Its leaves are bold and entire, and grow in pairs and opposite, its stamens and carpels are numerous. Beginning to flower in the early autumn, it lasts until winter commences. The designer should also make the acquaintance of the wild member of this family, *Clematis vitalba*. Its petals are four in number, small and are reflexed, the stamens or filaments are long, giving the name of *Old Man's Beard*, another and more common name being *Traveller's Joy*. The petals are whitish-green. It is to be found abundantly in the South of England. The Clematis is one of those plants that climb by means of their leaf stalks and not by tendrils.

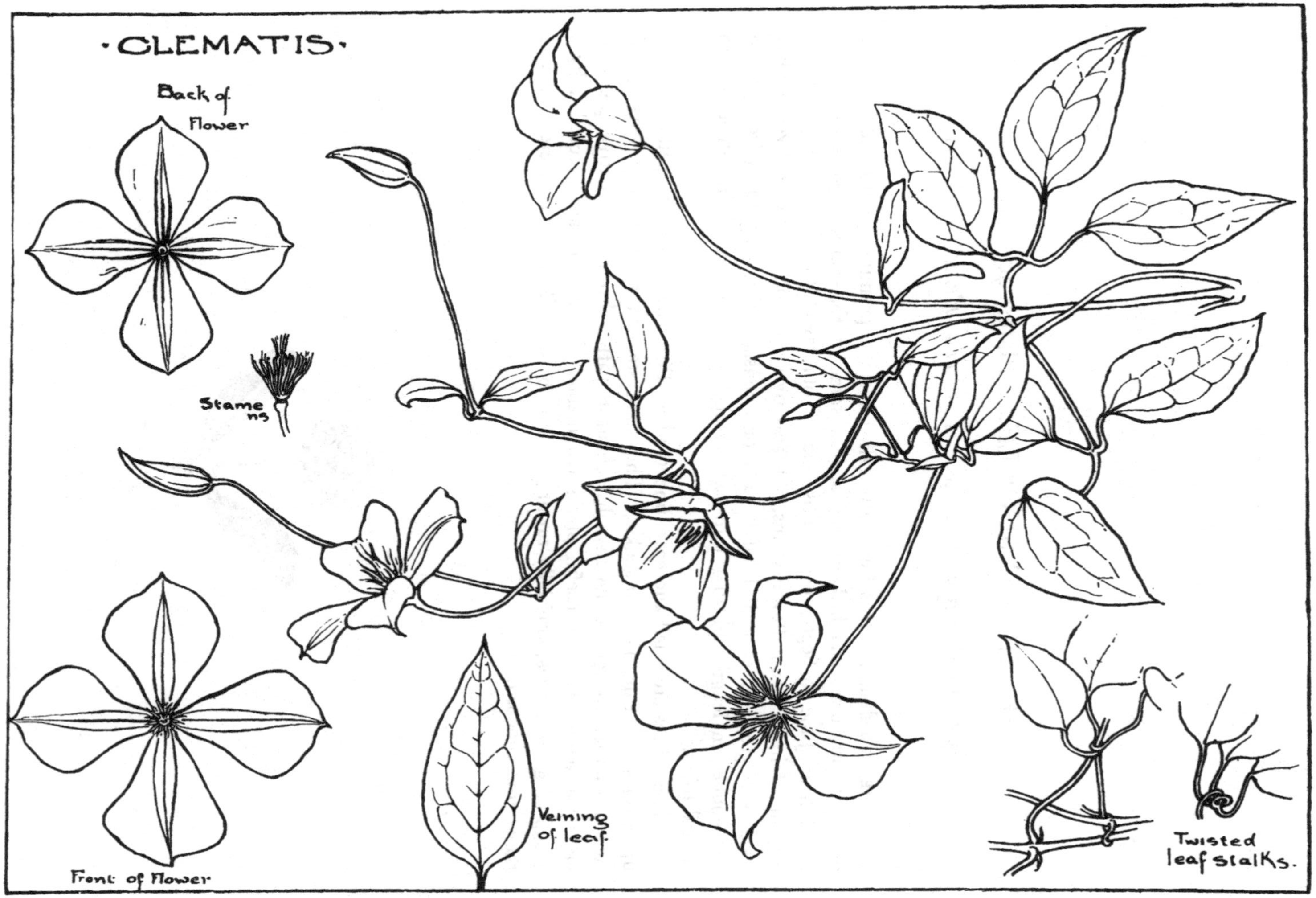

Plate 89.

·CLEMATIS·

Back of Flower

Stamens

Front of Flower

Veining of leaf

Twisted leaf stalks.

THE HONEYSUCKLE. Plates 90, 91, 92.

(*Nat. Ord. Caprifoliaceæ. Lonicera Periclymenum.*)

THE Woodbine, as it is frequently called, forms a prominent feature along some of our country lanes and hedgerows ; and in some of the southern counties may be found from June to September. It is one of our favourite climbers, not only on account of its beautiful flower forms, but also by reason of its strong, sweet odour. Its flowers grow in terminal heads ; each one consists of a long tube, from one to two inches in length, splitting into two petals, the larger one being divided into several lobes.

The calices are small and have fine teeth. The leaves grow in pairs, and the upper ones, as a rule, have no stalks. The main stem grows to a great length, and twines from left to right round any obstacle in its path. From a decorative point of view it is one of the most useful plants obtainable. Its flowers are succeeded by a group of green berries, gradually turning to a deep red. The stamens are five, and the stigma knobbed.

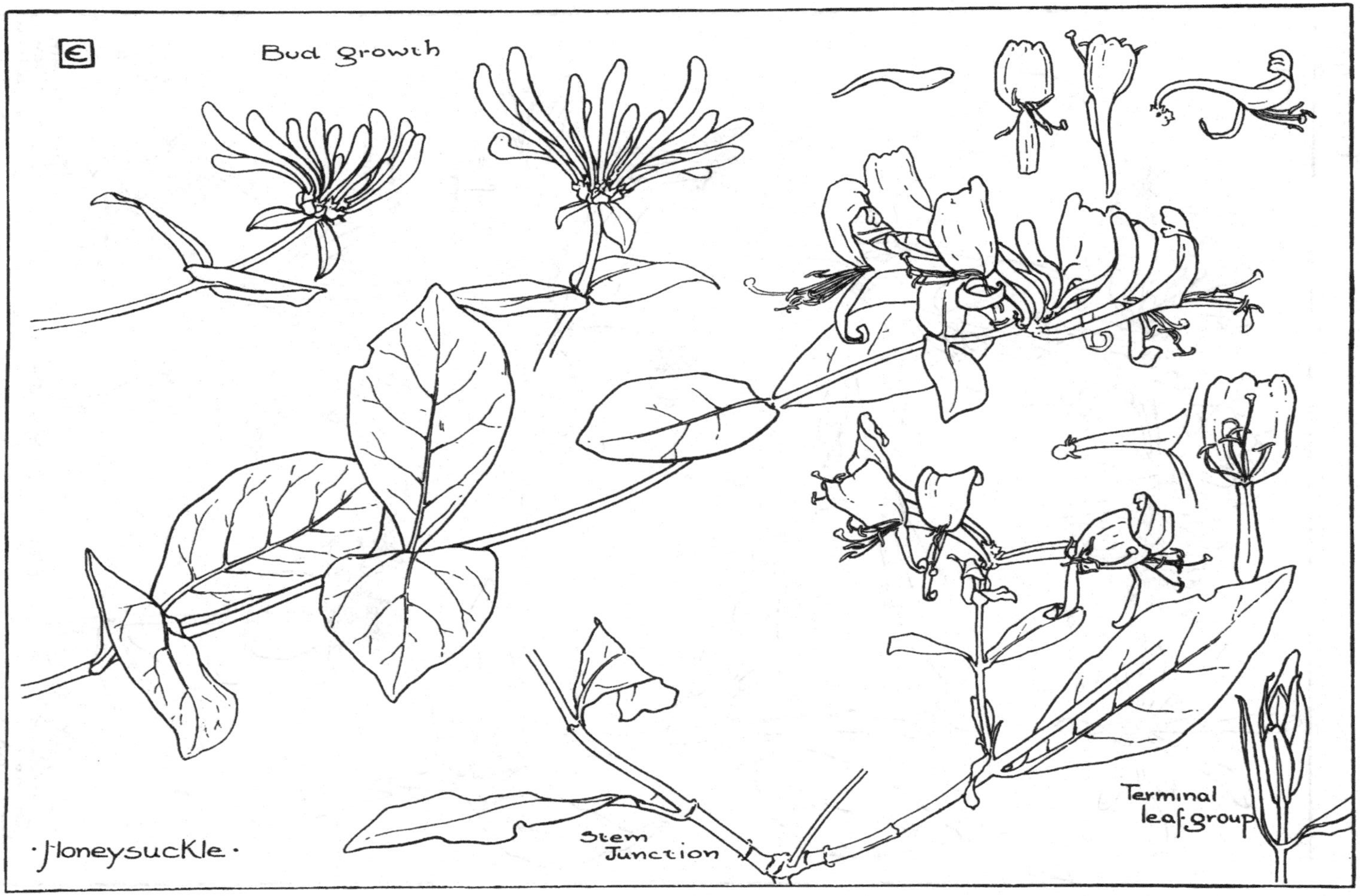

Plate 90.

Terminal
leaf group

Stem
Junction

Bud growth

·Honeysuckle·

Plate 91.

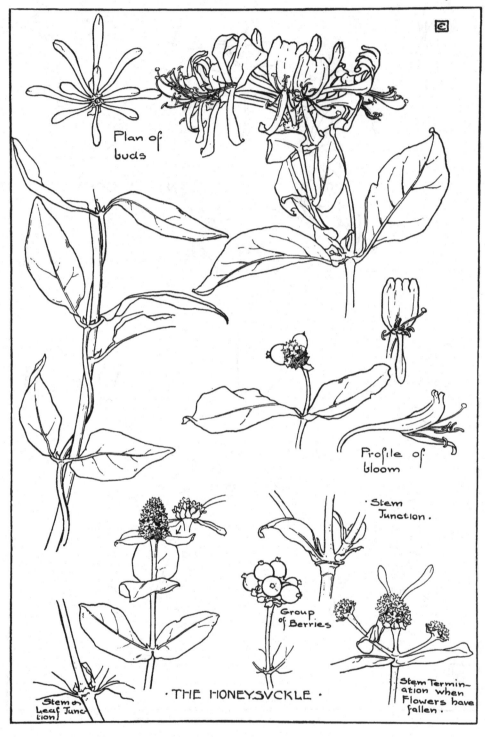

Plan of
buds

Profile of
bloom

Stem
Junction.

Group
of Berries

Stem or
Leaf Junc-
tion

· THE HONEYSVCKLE ·

Stem Termin-
ation when
Flowers have
fallen.

Plate 92.

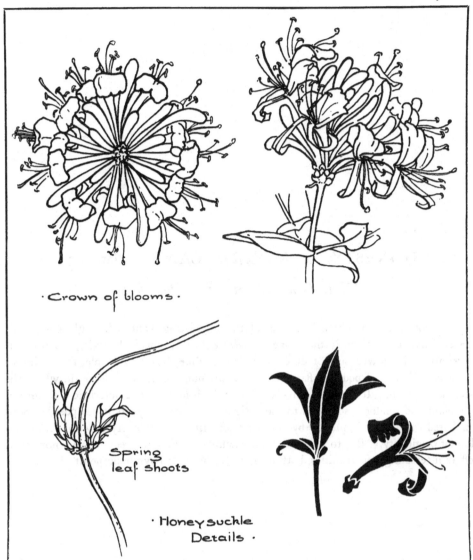

· Crown of blooms ·

Spring
leaf shoots

· Honeysuchle
Details ·

HONESTY, or MONEY-PLANT. Plate 93.

(*Nat. Ord. Cruciferæ. Lunaria.*)

THERE are two cultivated species of this interesting plant, whose glistening seed vessels have earned for it the name of *Money-flower*; one is biennial, the other is perennial. Its flowers are ot different colours, white, lilac and purple, from May to August. The seed-pods are flat and oval, containing a satin-like tissue, in which the seeds are hidden. It is these satin-like discs which form so characteristic a feature of the plant and render it so useful to the designer. Of recent years it has possibly been exploited rather too frequently by designers affecting a certain style, with a resulting loss of originality. But to a designer approaching it without preconceived notions it offers interesting opportunities, both from an æsthetic and symbolic point of view.

Plate 93.

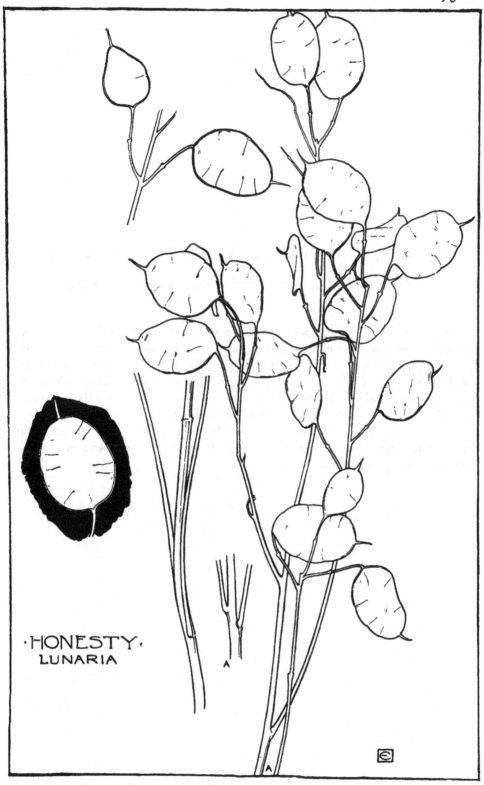

·HONESTY·
LUNARIA

EASTER, OR BERMUDA, LILY. Plates 94, 95.

(*Nat. Ord. Liliaceæ. Lilium longiflorum Harrisi.*)

THIS beautiful Lily is so well known to designers and floral decorators that a description seems superfluous ; but as this might also be said or many other flowers figuring here it is not a sufficient excuse for omitting it. The flower is a perianth, and is terminal. It has six oblong and pointed segments, creamy-white with distinct parallel markings, terminating in a tube. The stamens are six in number, the anthers of which are a deep yellow, the pistil is long and has a three-lobed stigma. The leaves are lanceolate and sessile, with parallel veining, and grow spirally round the stem. The main stem is branched and round in section. Flowers in spring and summer.

Plate 94.

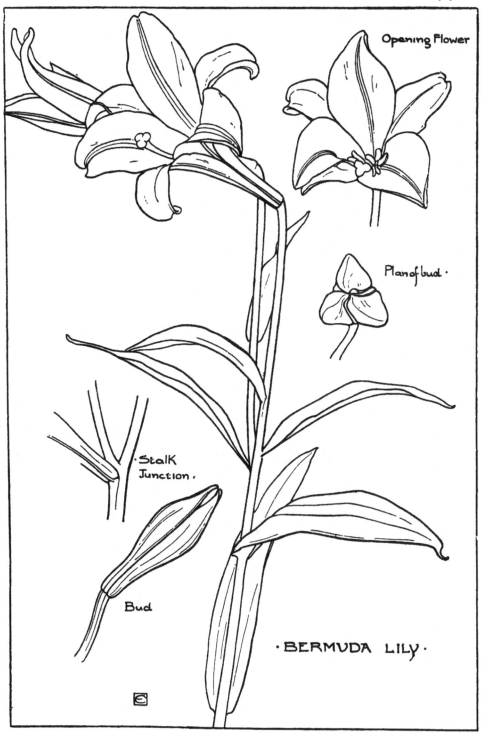

Opening Flower

Plan of bud.

·Stalk Junction·

Bud

·BERMVDA LILY·

Plate 95.

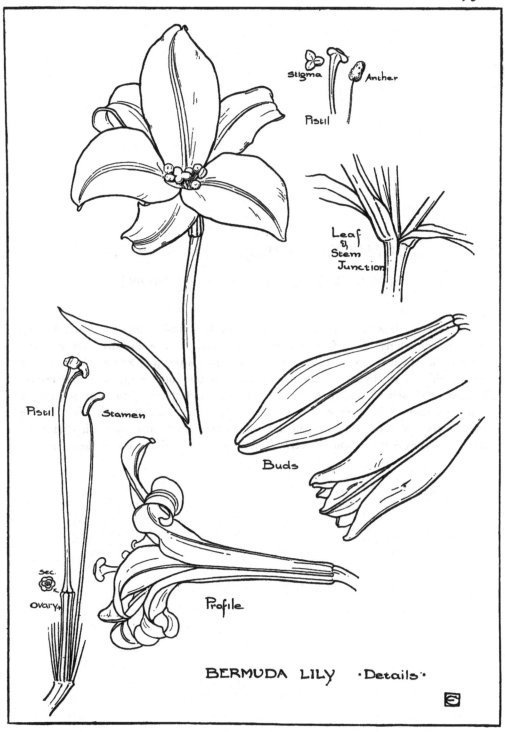

Stigma
Anther
Pistil

Leaf & Stem Junction

Pistil   Stamen

Buds

Sec.

Ovary

Profile

BERMUDA LILY   ·Details·

COMMON HOLLY. Plates 96, 97.

(*Nat. Ord. Ilicineæ. Ilex Aquifolium.*)

THE Holly is well known as an erect much-branched evergreen shrub, bearing in the spring whitish-green flowers of four petals, and calyx of four divisions, growing in thick clusters in the axils of the leaves. The fruit is the well-known red or yellow berry, giving a vivid touch of colour in contrast to the sombre green of the leaves. The leaves are shortly stalked, ovate in form, thick and shining; they are sometimes entire, sometimes with wavy outlines and strong, sharply-pointed coarse teeth. The stamens are also four in number, inserted on to the corolla and alternating with the petals. Berries Autumn and Winter. The leaves grow spiral-wise on the stem.

Plate 96.

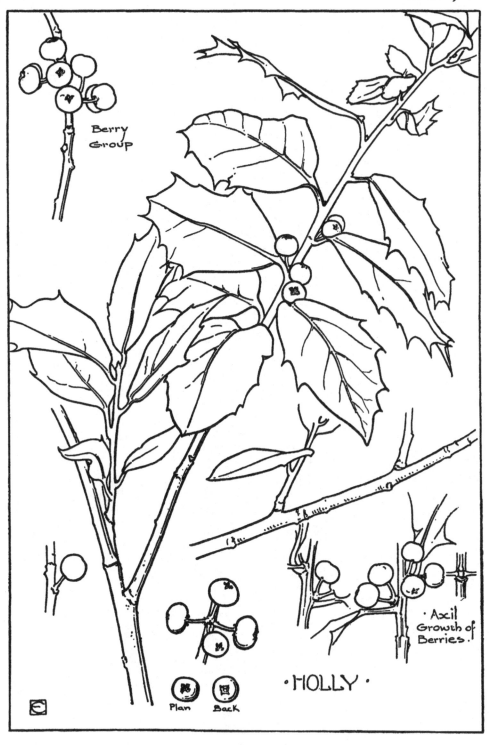

Berry
Group

Plan   Back

· Axil
Growth of
Berries.

· HOLLY ·

Plate 97.

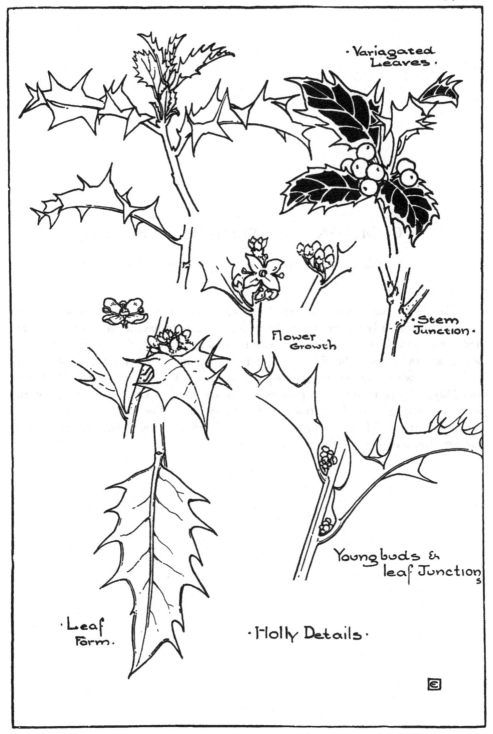

Variagated Leaves.

Flower Growth

Stem Junction.

Youngbuds & leaf Junction

Leaf Form.

· Holly Details ·

COMMON MISTLETOE. Plate 98.

(Nat. Ord. Loranthaceæ. Viscum album.)

THIS is the only British species of this order, and is to be found on a variety of trees, but particularly on the Apple, for which it seems to have a decided predilection, and rarely on the Oak. The stem is green and smooth, with bonelike joints. The leaves, which are thick and yellowish, grow in pairs opposite. It is most familiar to us in the winter, when bearing its whitish-green berries. The flowers appear in March and May, but are rather inconspicuous; they are nearly sessile in the forks of the branches, having four short, thick, triangular petals. The male flowers grow in groups of three and five, in a sort of cup-shaped bract. The female usually solitary, but sometimes two or three together.

Plate 98.

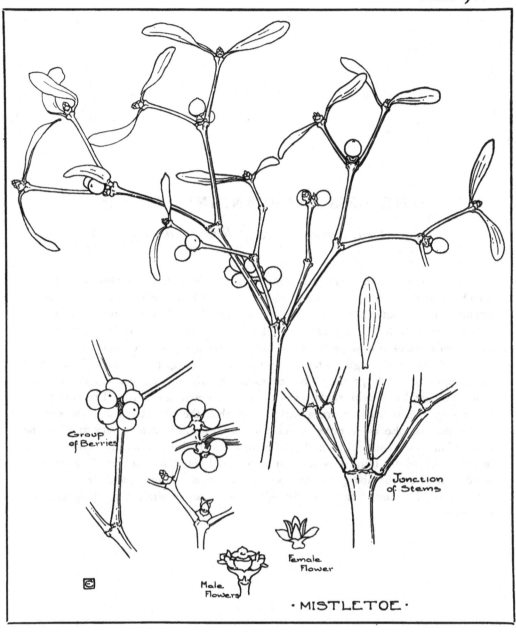

Group of Berries

Junction of Stems

Male Flowers

Female Flower

· MISTLETOE ·

THE LESSER CELANDINE. Plate 99.

(*Nat. Ord. Ranunculaceæ. Ranunculus Ficaria.*)

" SPRING is coming, thou art come," writes Wordsworth of this little flower, which accurately describes its early appearance. There are few more welcome or brighter blooms that usher in the early spring than the little Celandine coming " ere a leaf is on a bush." Its golden star-like blossom lights up the hedge banks and moist shady places in most pastures and lanes. Belonging to the Buttercup order, it consists of from six or eight to ten or more bright yellow glossy petals, and a calyx of three sepals of a pale green. The leaves are mostly radical, and are cordate and sometimes angular in form, slightly thick, and dark green in colour. The flowers are borne singly, on stems that are a little longer than the leaf stalks and are slightly channelled, and also taper rather quickly to the flower, broadening very much at the root. The root consists of a number of oblong or cylindrical tubers and fibres. The height of the plant is from three to eight inches. Although looked upon as a common weed, it inspired the above quoted poet to sing its praises often, and a copy of it is carved upon his tomb. It might fittingly serve as an inspiration to designers and craftsmen.

Plate 99.

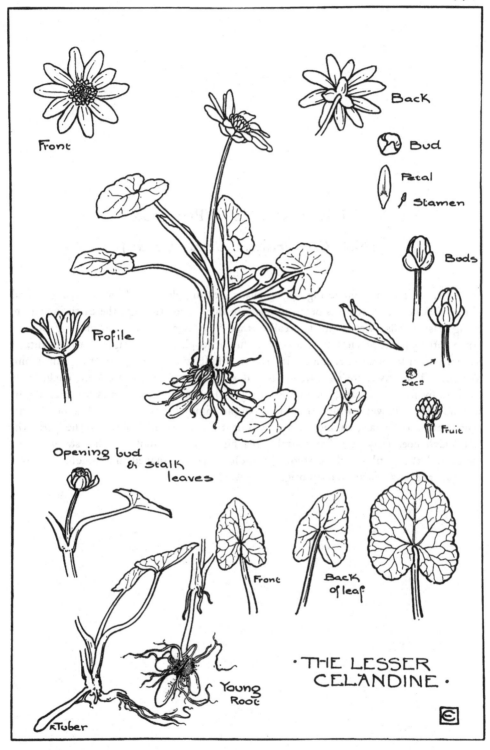

Front

Back

Bud

Petal

Stamen

Profile

Buds

Seed

Fruit

Opening bud
& stalk
leaves

Front

Back
of leaf

Young
Root

·THE LESSER
CELANDINE·

R.Tuber

THE CROCUS.   Plate 100.

(Nat. Ord. Irideæ.   Crocus vernus.)

THIS is an early-flowering spring plant, bearing flowers of purple, yellow, and white. When the Crocuses begin to push up their heads through the earth in spite of biting east winds, we begin to realise that "Springe ys i-comen in." From an ornamentist's point of view it is a very useful little plant, and might with advantage be more often turned to account. Its root is a bulb with a scaly covering and white rootlets. The leaves are narrow, linear and radical, dark green in colour, enclosed at the base by fibrous sheaths, they grow nearly as high as the flower and are angular in section. The flower, which in the example figured was purple, has nearly six equal petals or sepals, ending in a very long tube the same colour as the perianth, and which does duty as a flower-stalk, reaching to the root. The stamens are three, and the pistil has three stigmas, which are wedge-shaped and notched. Both stamens and stigmas are a deep bright yellow.

Plate 100.

Sec?

· CROCUS ·

A SHORT GLOSSARY OF
BOTANICAL TERMS.

ACICULAR, needle-shaped, *see* fig. 10, pl. 2.

AMPLEXICAUL, embracing or clasping the stem, as the leaf in fig. 26, pl. 2.

ANTHERS, the heads of the stamens which contain the pollen, usually lobed.

AXIL, the angle formed by the leaf and stem of a plant.

BLADE, the broad part of a leaf as distinct from the stalk or petiole.

BRACT, irregularly developed leaves, generally found at the base of the flower-stalk. They are, as a rule, different in form and colour from the stalk leaves. *See* the bracts on the Bluebell.

BERRY, one form of fruit. The fruit of a plant is formed by the pistil or, usually, that part of it called the *ovary*, which does not ripen until the petals have fallen.

CALYX, the outer whorl of leaves of a flower or bud, usually green and smaller than the petals, sometimes very minute. It is made up of the sepals.

CAPSULE, a dry, hollow seed-vessel, as in the Poppy.

CAPILLARY, hair-like.

CAPITULUM, a form of inflorescence ; a *head* of flowers without pedicels or stalks, as in such flowers as the Cornflower or Dandelion.

CARPEL, the divisions of the ovary or seed-vessel.

COROLLA, the petals of a flower. They are either separate or united.

CONE, is a hard fruit of the Pine or Fir tree, consisting of scaly bracts which shelter the seed.

CONNATE, is applied to opposite leaves which clasp the stem by being united at their bases.

CORDATE, heart-shaped, as the Violet leaf, *see* fig. 11, pl. 2.

CORYMB, a form of inflorescence, arranged as a raceme, but with the lowest flower stalks longest, giving a flat top to the group.

CRENATE, notched round the edge with rounded teeth, *see* fig. 2, pl. 2.

CATKIN, a close spike of small flowers and bracts, which hangs loosely, as in the Hazel and Alder.

CLAW, the narrow base of a petal ; the broad upper part is called the *limb*.

CRUCIFEROUS, cruciform, with four petals arranged in the manner of a cross.

CYME, an inflorescence which resembles the corymb, but develops from the centre —*i.e.* is centrifugal, as the Golden Saxifrage.

DECUSSATE, referring to leaves which grow in opposite pairs, crossing each other at right angles.

DENTATE, tooth-like divisions round the margin.

DRUPE, a nut enclosed in a fleshy covering, as the Plum or Cherry.

DRUPEL, the small drupes which compose the Blackberry or Raspberry.

ENTIRE, applied to leaves having undivided margins.

EPICALYX, the outer calyx or involucre, consisting of bracts or scales.

FASCICLE, resembles an *umbel*, but with stalks of different lengths.

FILAMENT, the thread-like portion of the stamen supporting the anther.

FLORET, a small flower. Composite flowers are made up of florets, which are usually called petals.

FRINGED, having a border of fine hair-like projections, *see* fig. 9, pl. 2.

FRUIT, the ripened seed-vessel of a plant, with its contents and covering.

GLABROUS, without hairs, smooth.

GLAUCOUS, covered with a bloom of pale bluish-green, like the leaves in the Opium Poppy.

GENUS (GENERA), a division of a natural order, comprising allied species and varieties.

HASTATE, a leaf form so-called from its resemblance to a halberd, *see* fig. 20, pl. 2.

HYBRID, a cross between two varieties, *e.g.* the Shirley Poppy.

IMBRICATION, the overlapping of scales or bracts, like roof tiles. See the scales that protect the leaf-buds of the Horse-chestnut or the Ash.

INFLORESCENCE, a term used to denote the arrangement of flowers on the stem.

INVOLUCRE, an arrangement of bracts or scales, calyx-like in form, enclosing many florets, as in the Daisy or Dandelion.

LANCEOLATE, the shape of a leaf when it is much longer than it is broad, broadest below the middle, tapering towards the summit like a lance head.

LEAFLET, the separate divisions of a compound leaf, when divided to the stalk or petiole, *e.g.* the Dog-rose leaf.

LEGUME a form of fruit, produced by the plants in the order Leguminosæ ; *e.g.* Peas and Beans.

LIGULATE, strap-shaped.

LINEAR, long and narrow, as the leaves of Grasses, *see* fig. 15, pl. 2.

LOBE, the division of a leaf, when such division does not reach the midrib.

MEMBRANOUS, a thin fibrous substance.

MONOPETALOUS, when the petals are united, as in the Convolvulus.

NATURAL ORDER, the name under which *genera* resembling each other are classed. It has the same meaning as family.

NET-VEINED, when the veins of a leaf are spread and connected with each other, forming a network. Reticulated. This is one of the principal distinctions between leaves; the parallel-veined leaves, like the leaf of the *Solomon's Seal*, form the next greatest variety, as distinguished by the veining.

NODE, that part of the stem whence the leaves spring. An *internode* is the part of the stem between two nodes.

NUT, one form of fruit, hard and dry. A one-seeded carpel, *e.g.* the Acorn and Hazel nut. A *samara* is a winged nut.

OBCORDATE, the form of a leaf which is inverse heart-shaped, *see* fig. 12, pl. 2.

OBLIQUE LEAVES, leaves with unequal sides, *see* fig. 22, pl. 2.

OBOVATE, the same as ovate, but with the broadest part of leaf a little above the middle.

ORBICULAR, of circular form and flat, applied to rounded leaves, *see* fig. 2, pl. 2.

OVATE, egg-shaped, with the broad end downwards, *see* fig. 6, pl. 2.

OVARY, situated at the base of the pistil, contains the rudiments of the future seeds. It becomes the fruit of a plant.

PALMATE, shaped like the hand, digitate, diverging from one point like the Horse-chestnut leaves.

PANICLE, a form of inflorescence, like the *raceme*, but branched.

PAPILIONACEOUS, butterfly-shaped, *see* Sweet Pea.

PARASITE, a plant which derives its nourishment from another, as the Mistletoe.

PEDICEL, the stalk of a single flower.

PEDUNCLE, a flower-stalk, sometimes only supporting a single flower, more generally the main floral axis.

PELTATE, a flat form of leaf, like the Nasturtium, when the petiole or leaf stalk is attached to the under-surface, and from which point the veins radiate.

PEPO, is a fruit form produced by the order Cucurbitaceæ, as the Marrow.

PERFOLIATE, a form of leaf attachment, when the base of the leaf not only clasps the stem, but passes round it, so that the stem appears to pierce the leaf.

PERIANTH, generally applied to the floral envelope, when it is not differentiated into calyx and corolla. When either the calyx or corolla is absent.

PERICARP, the outer covering of the fruit or seed.

PETIOLE, the leaf-stalk, as distinct from the *pedicel*, the flower-stalk.

PINNATE, feathered, the form of some compound leaves, when the lobes or leaflets are arranged regularly on either side of a midrib, as in the Wild Rose, or the Ash.

PISTIL, the centre form or organ of a flower, usually consists of a delicate column of three parts—the ovary, the style, and the stigma.

POD, a dry fruit, a one-celled, two-valved, many-seeded vessel. Applied to the fruits of the Pea and Bean tribe. A legume.

POME, a fleshy seed-vessel, such as the Apple and Pear.

POUCH, sometimes used to denote a short seed-pod or vessel, such as that of the Honesty.

RACEME, a form of inflorescence ; when the flowers are arranged in the manner of a spike, but each having a separate stalk or pedicel direct from the main axis. If the pedicel is branched, the arrangement becomes a panicle.

RADICAL LEAVES, growing straight from the root, at the base of the flower-stalks.

RAY, a term applied to the outer florets of some of the Composite flowers, as the Daisy.

RECEPTACLE, that part of the flower which supports the other parts ; well seen in the Dandelion details.

RENIFORM, kidney-shaped, *see* fig. 13, pl. 2.

RHIZOME, a root-like stem or root stock, often underground, putting forth roots from the under-side.

RUNCINATE, deeply toothed at the margin, the teeth pointing to the root, as in the Dandelion, *see* fig. 23, pl. 2.

SAGITTATE, shaped like an arrow, *see* fig. 17, pl. 2.

SAMARA, a winged fruit, *see* fruit of Sycamore.

SCAPE, a long leafless stem, straight from the root, carrying a flower.

SEGMENT, the division of a leaf or flower.

SEPAL, a segment of the calyx.

SESSILE, applied to flowers and leaves without stalks. A *stigma* is sessile when it has no *style*, as in the poppy, an *anther* when it has no *filament*.

SERRATE, notched or toothed like a saw, as in the leaf of a Wild Rose, *see* fig. 3, pl. 2.

SHEATH, the lower part of a leaf or its stalk when it is rolled round the stem.

SILIQUE, a long pod, such as those borne by the Cruciferæ, or of the Pea.

SILICLE, a short or broad pod.

SIMPLE, when the blade of a leaf is of one piece, the reverse of compound, *see* fig. 1, pl. 2.

SPADIX, a succulent spike, bearing small, sessile flowers, enclosed in a spathe, as in the Cuckoo Pint.

SPATHE, a sheath-like bract or floral leaf, enclosing the inflorescence, as in the Cuckoo Pint.

SPATULATE, applied to a leaf when the top part is short and broad, and the lower part long and tapering, *see* fig. 4, pl. 2.

SPIKE, a form of inflorescence, when the blossoms are numerous and sessile, arranged on a central axis or stem.

STAMEN, a number of delicate organs within the corolla or centre of flower, consisting of the anther and filament.

STANDARD, the large upright petal of a papilionaceous flower, *e.g.* the Pea.

STIGMA, the summit of the pistil, receiving the pollen.

STIPULE, lateral appendages, or small leaf forms at the base of the leaf-stalk.

TAP-ROOT, when the main root descends perpendicularly into the earth, sending off only fine fibers.

THYRSUS, an inflorescence similar to a panicle, but the flowers set close to the central axis by more solid pedicels, as in the thyrsus of the Horse-chestnut.

TERMINAL, ending a stalk, proceeding from the end, as in the Ox-eye Daisy.

TUBEROUS, thick fleshy roots or tubers.

UMBEL, an inflorescence, when the pedicels apparently start from the same point, and are nearly all the same length.

VENATION, applied to the arrangement of the veins in the leaves.

WHORL, leaves, branches, or flowers, when arranged in a circular manner round the stem or central axis.

BRADBURY, AGNEW & CO., LD., PRINTERS, LONDON AND TONBRIDGE.